Butchery & Sausage-Making

FOR

DUMMIES®

by Tia Harrison

WILEY

John Wiley & Sons Canada, Ltd.

Butchery & Sausage-Making For Dummies®

Published by
John Wiley & Sons Canada, Ltd.
6045 Freemont Blvd.
Mississauga, ON L5R 4J3
www.wiley.com

For general information on John Wiley & Sons Canada, Ltd., including all books published by John Wiley & Sons, Inc., please call our distribution centre at 1-800-567-4797. For reseller information, including discounts and premium sales, please call our sales department at 416-646-7992. For press review copies, author interviews, or other publicity information, please contact our publicity department, Tel. 416-646-4582, Fax 416-236-4448.

For technical support, please visit www.wiley.com/techsupport.

Wiley publishes in a variety of print and electronic formats and by print-on-demand. Some material included with standard print versions of this book may not be included in e-books or in print-on-demand. If this book refers to media such as a CD or DVD that is not included in the version you purchased, you may download this material at http://booksupport.wiley.com. For more information about Wiley products, visit www.wiley.com.

Library and Archives Canada Cataloguing in Publication

Harrison, Tia

    Butchery & sausage-making for dummies / Tia Harrison.

Includes index.

Issued also in electronic format.

ISBN 978-1-118-37494-8

    1. Meat cutting. 2. Sausages. 3. Cooking (Sausages).

4. Meat—Preservation. I. Title.

TS1962.H37 2012   664'.902   C2012-902751-0

978-1-118-38743-6 (ebk); 978-1-118-38744-3 (ebk); 978-1-118-38745-0 (ebk)

Printed in the United States

SKY10030734_102121

WILEY

Butchery & Sausage-Making

FOR

DUMMIES®

About the Author

Tia Harrison is co-founder of the Butcher's Guild; executive chef and co-owner of Sociale, a northern-Italian–inspired restaurant; and co-owner of Avedano's Holly Park Market, a neighborhood butcher shop that focuses on whole-animal butchery.

Since 2003, Tia has been Sociale's executive chef, where she has worked to expand her expertise as a food professional in San Francisco. Located in the Presidio Heights district of San Francisco, Sociale has been an integral part of the "farm to table" approach to dining in the Bay Area and proudly serves only the best in locally grown produce, sustainably farmed meats, and wild-caught seafood.

In 2007, Tia Harrison opened Avedano's Holly Park Market, the first women-owned and women-run butcher shop in the United States. Inspired to revive the dying art of butchering by hand and to support small farms and sustainable food systems, Avedano's and its owners are leaders in the butcher's revolution, focusing on educating others in the art of butchering and supporting consumers, local farmers, and the community in all things that are meaty and delicious. Tia Harrison is focused on education, community building, and good food.

Tia, along with her friend and business partner Marissa Guggiana, founded the Butcher's Guild in 2010. The Guild, a fraternity of meat professionals bound to creating a support system for the butchery industry, focuses on education and camaraderie to achieve a communal goal: a network of successful, skilled, independent butchers all across the country.

Dedication

For Dacia. You are the light of my life.

Author's Acknowledgments

To butchers, chefs, salumists: You inspire me. The meat industry has given me so much. A big, huge "thank you" to all the passionate people who work to support the craft of butchery. Thank you to *all* the members of the Butcher's Guild. Your friendship and support mean so much to me. I am truly honored to know all of you. Thank you to Marshka Kiera for literally rocking our world and being such a kind, wonderful, and helpful friend to the Guild.

To my family: Ellice, Matt, Nya, Rocco, Ben, Carly, Gary, Marianne, Nathaniel, Travis, Caitlin, Ellen, Paul, Flora. All of you have helped me in my career. Thank you for always listening to my wild, crazy, entrepreneurial schemes (and pretending to be excited). I appreciate all your love and support. To BB, ily so much; thank you.

To all of the inspiring people who have contributed to and helped me with this book: Christopher Arentz, David Budworth, Bryan Butler, Craig Deihl, Mark DeNittis, Josh Donald, Brad Farmerie, Marissa Guggiana, Peter Hertzmann, Matthew Jennings, Josh Martin, Julio Mis, Pepe, Stephen Pocock, Berlin Reed, Gregg Rentfrow, Adam Tiberio, Kari Underly, Sarah Weiner, and Oscar Yedra.

Extra special thanks go to Adam Tiberio, Bryan Butler, and Julio Mis. Adam, thanks for sharing your knowledge, skills, and passion with me; for the advice and information you've provided; and for being an educator. Bryan, thank you for your help with this project; your feedback and support have been huge. Julio, thank you for always being on my team, no matter how much I ask for. Working with you is such a comfort to me. Thank you for all of your help with this book. You have done so much.

To my dear, dear friend and business partner, Marissa Guggiana. You hold a very special (almost creepy, perhaps stalker-y) place in my heart. You inspire me and give me the confidence to do impossible things. Your help during this experience was immeasurable. I truly could not have done it without you.

To David Nichol, you are a great friend and business partner. Thank you for understanding me and always letting me follow my dreams. Sociale is where my heart is.

To Angela Wilson and Melanie Eisemann: Thank you for believing in me, always being there for me, and for being such good friends and good people. You have both inspired me.

Publisher's Acknowledgments

We're proud of this book; please send us your comments at http://dummies.custhelp.com. For other comments, please contact our Customer Care Department within the U.S. at 877-762-2974, outside the U.S. at 317-572-3993, or fax 317-572-4002.

Some of the people who helped bring this book to market include the following:

Acquisitions, Editorial, and Vertical Websites

Acquisition Editor: Anam Ahmed

Project Editor: Tracy L. Barr

Production Editor: Pauline Ricablanca

Editorial Assistant: Kathy Deady

Technical Editor: Bryan Butler

Cover photos: © maria bacarella

Cartoons: Rich Tennant (www.the5thwave.com)

Composition Services

Senior Project Coordinator: Kristie Rees

Layout and Graphics: Carrie A. Cesavice, Jennifer Creasey, Brent Savage

Proofreaders: Melissa Cossell, Wordsmith Editorial

Indexer: Christine Karpeles

Recipe Tester: Emily Nolan

John Wiley & Sons Canada, Ltd.

Deborah Barton, Vice President and Director of Operations

Jennifer Smith, Publisher, Professional & Trade Division

Alison Maclean, Managing Editor, Professional & Trade Division

Publishing and Editorial for Consumer Dummies

Kathleen Nebenhaus, Vice President and Executive Publisher

David Palmer, Associate Publisher

Kristin Ferguson-Wagstaffe, Product Development Director

Publishing for Technology Dummies

Andy Cummings, Vice President and Publisher

Composition Services

Debbie Stailey, Director of Composition Services

Contents at a Glance

Recipes at a Glance

Table of Contents

Part III: Pork Butchery 121

Chapter 8: Porky Pig: Understanding the Beast 123

Chapter 9: Pork: Cutting It Up 133

Part V: Sausage-Making and Using the Whole Animal ...245

Introduction

• •

Sixty years ago, the work of butchers seemed a lot less mysterious. I remember going to a butcher shop in Alameda, California, with my grandmother when I was very young. While my grandma shopped for dinner, the butcher's long arm reached over the counter and handed me a slice of paper-thin, house-made bologna. I remember thoroughly enjoying the meat snack, even though I thought I didn't like bologna. It was the gesture that made it taste so great, and even at the time, it seemed to me a very sweet and special outing.

I still love butchery. As a chef, learning to butcher was just a part of the natural progression of my craft. I wanted to know more about the meat I was serving in my restaurant; I knew how to sear a steak or braise a short rib, but what I really wanted to know how was how to cut it. I have so much respect for butchers. When we opened Avedano's in 2007, making the transition from chef to butcher was an eye-opener to say the least. I thought it would be a lot easier than it was. I had a lot to learn. I have been fortunate enough in my career to know many great, talented people (many of whom have contributed to this book) that have shared their meat expertise and knowledge with me. We need butchers, we need their skill, we need their perspective, and we need to support them to make sure that they stick around. They are a very important part of a healthy, local food system. I raise my knife to them, their heritage, and their craft.

About This Book

No butchery book — or at least no butchery book that you can carry around without a hydraulic lift — can tell you everything there is to know about how to cut meat. In this book, I don't even try. Instead, I tell you what you need to know to begin your butchery adventure and to do so safely and successfully.

In this book, you can easily find information, like

- ✔ The proper way to hold and use the knives you'll need for butchery
- ✔ Instructions for cutting a chicken into eight pieces or how to french bone-in chops
- ✔ The difference between butchering on a table or on the rail

- ✔ Complete instructions for dietary staples, like poultry, beef, and pork butchery, as well as instructions for animals that are less common in the typical North American diets

- ✔ Sausage-making instructions and many delicious and easy-to-make recipes

The great thing about this book is that *you* decide where to start and what to read. It's a reference you can jump into and out of at will. Just head to the table of contents or the index to find the information you want.

Conventions Used in This Book

To help you navigate through this book, I've set up a few conventions:

- ✔ *Italic* is used for emphasis and to highlight new words or terms that are defined.

- ✔ **Boldfaced** text is used to indicate the action part of numbered steps.

- ✔ `Monofont` is used for Web addresses.

- ✔ Recipes show measurements in grams (a common convention among butchers), but I've added equivalents for those who don't have gram scales.

What You're Not to Read

I've written this book so that you can find information easily and easily understand what you find. And although I'd like to believe that you want to pore over every last word between the two yellow covers, I actually make it easy for you to identify "skippable" material. This is the stuff that, although interesting and related to the topic at hand, isn't essential for you to know:

- ✔ **Text in sidebars:** The sidebars are the shaded boxes that appear here and there. They share personal stories and observations, but aren't necessary reading.

- ✔ **Anything with a Technical Stuff icon attached:** This information is interesting but not critical to your understanding of butchery.

Foolish Assumptions

Every book is written with a particular reader in mind, and this one is no different. As I wrote this book, I made a few assumptions about you:

- ✔ You care about the quality of the food you eat and supporting local food systems.

- ✔ You believe that age-old crafts like butchery are worth learning because they're timeless and vital to healthy diets and supporting your community.

- ✔ You're interested in getting back to basics, and butchering your own meat is one of the ways you've decided to achieve this goal.

- ✔ You're new to butchery — in fact, you may have done little more with a knife that slice cheese or carve/hack your Thanksgiving turkey — but are interested in knowing how to use knives (and other cutting implements) competently and safely.

- ✔ You're passionate about using the whole animal.

How This Book Is Organized

To help you find information that you're looking for, this book is divided into six parts. Each part covers a particular aspect on butchery and contains chapters relating to that topic.

Part 1: Time to Meet Your Meat!

This section is chock-full of all the basics you need to know about butchery. It includes information about the history of butchery and meat curing and the important role that butchers play in our food system. I tell you how to shop for meat, how to read labels, identify meat cuts, and substitute for meal planning. I share with you a quick overview of meat science and aging meat, and I also cover how to use knives properly and, most importantly, how to use them safely.

Part II: Poultry, Rabbit, and Lamb Butchery

When learning a new craft or skill, we all start somewhere. If you want to learn to butcher, this is where you should start. Begin with poultry and work your way up to lamb. Butchering for the first time is not an easy task, but with the right practice and progression, you'll gain the skills that will put you on the path to proficient meat-cutting prowess.

Part III: Pork Butchery

In this part, I identify primals, subprimals, and retail cuts for pork and explain how to butcher a whole hog. You can butcher a pig in many different ways, and the process I show you here is just one of them, but it's a process you should be able to follow with relative ease after you're comfortable butchering the smaller animals in Part II. Not only will you end up with pounds of wonderful pork cuts to enjoy for a number of meals, but, as an added bonus, you can use the cuts produced in this part for decadent sausage-making and further processing later, the topic of Part V.

Part IV: Beef Butchery

Beef butchery is, without a doubt, the most difficult and challenging of all the butchery tasks. In this part, I explain one method of breaking beef that is particularly exciting to me. Here, in addition to getting familiar with beef primals, subprimals, and a number of retail cuts, you also discover how to butcher a hanging carcass, as well as how to finish up the cuts on the bench.

Part V: Sausage-Making and Using the Whole Animal

Sausage-making is as much a part of butchery as cutting. As a butcher, you should have a plan for using the whole carcass, and further processing, which includes sausage-making, curing, and more, is the way to go. In this part, I tell you how you can make delicious sausages at home. I also give you a primer in meat curing, along with other delicious alternatives to preserving meat at home with culinary flare.

Part VI: The Part of Tens

Want to know the most common mistake new butchers make and how to avoid them? Curious about which cuts are best for grilling? Eager to hear secrets to making the best sausage? You can find the answers in this part.

Icons Used in This Book

The icons in this book help you find particular kinds of information that may be of use to you:

You see this icon anywhere I offer a suggestion or bit of advice — like how to save time or an easier way to make a particular cut — that can help you with the task at hand.

This icon points out important information that you want to remember.

In butchery, you're using very sharp knifes and working in an environment where your hands will often be wet, slippery, and cold. This icon highlights mistakes that can get you into trouble and offers important rules, guidelines, and advice that can help you avoid injuring yourself.

This icon appears beside information that is interesting but not necessary to know. In fact, feel free to skip the info here if you want. Doing so won't impair your butchering ability.

The Butcher's Guild is a fraternity of meat professionals focused on whole-animal butchery. Throughout this book, Butcher's Guild members have shared their butchery knowledge and skill with you via tips, industry information, and special techniques. Just look for this icon to get this insider information.

Where to Go from Here

This book is designed so that you can jump in and jump out as you like. If you want to know how to cut up the chicken you plan to have for dinner, for example, head to Chapter 5 for complete instructions. If you've ordered a forequarter of beef and have set aside your Saturday to butcher it, Chapter 12 has the info you need.

However, if you're new to butchery and need the very basics, start in Part I. Here you can find information on important terminology, kinds of cuts, necessary equipment, and safety instructions. Beyond that, decide which type of animal you want to butcher and go to the part devoted to that kind of butchery.

You can also use the table of contents to find broad categories of information or the index to look up more specific things.

Part I
Time to Meet Your Meat!

The 5th Wave   By Rich Tennant

©RICHTENNANT

"You can't really call yourself a butcher without a working knowledge of the bovine anatomy. Okay Vicky, we're ready for you."

In this part . . .

*I*n this part, you discover vital background information and nerdy meat facts that will set you on the right path to butchering your meat at home. As with any skill, you need to start with the basics, and quite often the basics are information about the history and inner workings of the trade.

This part also includes a lot of must-know info, too, like where to source your meat, what kind of equipment you need, and how to butcher efficiently and safely.

Chapter 1

The Butchery Room

In This Chapter
▶ Looking at the butcher's role
▶ Identifying the benefits of butchering your own meat
▶ Delving into the joys of sausage-making and other preservation techniques

What do butchers do? The obvious answer is that they cut meat, prepare it to be sold, and sell it. The less obvious answer requires thinking beyond the meat counter. Butchers are a powerful fulcrum in a healthy food system. Throughout recorded history, butchers have fed their communities, maintained a pivotal position in the marketplace, and even played a key role in local politics.

As a home cook who's interested in butchery, you are in a unique position to do the same as professional butchers, just on a smaller scale. Where professional butchers prepare meat for sale, you prepare meat to serve. Where professionals feed their communities, you feed your family. Where the pros hold pivotal positions in the market and key roles in local politics, you have significant influence not only in what your family buys, but also what kind of value your family places on having a reliable, healthy, local food supply. And if all politics is local and personal, you can't get more local — or personal — than what you prepare in your own kitchen or serve at your own dinner table.

In this chapter, I introduce you to butchers and the basics of butchery. I also explain why learning to butcher meat for your own family can produce benefits that stretch from your own dinner table to your community and beyond.

Industrialized meat

The centralization of the meat industry has fundamentally changed the art of butchery. Centralization is what happens when any organizations' activities become focused within a particular region or group. At the heart of this dysfunctional relationship between centralized/industrial meat and the craft butcher is the continuous demoting of skilled labor in order to make the butchery process more streamlined, inexpensive, and controllable.

With powerful efficiency and the ability to sustain fast growth, the commoditized meat industry has become a formidable feeding machine, supplying millions of eaters with boxed meat and value-added products regardless of the season or supply.

When business becomes more centralized, diversity and community become diluted. Yes, more people have access to consistent products, but regional differences slowly disappear and our connection to the process becomes equally faded. The competitive drive of these meat companies to maintain their concentrated share-hold of the marketplace is real. Mass farming practices like concentrated animal feeding operations (CAFOs) create farming environments that many scientists say are responsible for wide-spread pathogen propagation. The profits also become more centralized and local dollars are funneled away. When a community loses its financial independence and stability, it tends to lose its cultural vibrancy, as well.

Bottom line: We are what we eat, and dietary health epidemics have shown us that we can't make shortcuts with our diet and expect to circumvent the side effects.

Understanding the Importance of Ye Ol' Butcher Shoppe

When you get right down to it, a butcher's work is to prepare meat for sale. But that doesn't mean that butchers' responsibilities are limited to wrapping meat in cellophane and sticking a price tag on it. In fact, professional butchers perform a variety of tasks. Butchers, like other professionals, may also specialize in a particular area of butchery. I give you the details in the following sections.

Local butchers that work with whole animals are experts in meat cuts, meat sales, and the preparation of everything from steaks to sausages. The difference between butchers and meat cutters is that meat cutters are restricted by their environments to making only final cuts; they work in large grocery chains cutting prefabricated boxes of primal and subprimal cuts into steaks, roasts, and grinds. Everyone in this industry plays an important role, but in this chapter, I focus on the expert butcher.

Identifying what butchers do

Butchery is hard, laborious work. Lifting a carcass or sawing through thick bones takes a lot of body strength, but the final product — and the impact that butchers can have in both big (supporting local, sustainable food sources) and small (helping you choose the best cut for your family dinner) ways — is worth the hard work.

Selecting and preparing meat for sale

Here are the tasks that butchers perform to get meat ready for sale:

- **Selecting the carcass:** Industrialization and strict standardization of carcass condition have made selection irrelevant for the modern meat-cutter; but historically, it was always the role and responsibility of the butcher to know his stuff when it came to the overall quality of the animal. As local food systems are being re-created, this selection and exchange with the grower is regaining relevance.

- **Quartering or halving the carcass and then breaking it down into primary, or *primal*, cuts:** Primal cuts are the wholesale cuts into which an animal is first divided. How many primal cuts you have varies according to type of animal and the country you are in (different countries, and sometimes even regions within a country, often break a carcass down differently, according to local food traditions).

- **Deboning, trimming, or breaking primal cuts down further into steaks and roasts:** These are known as the *retail cuts*. They're what you see in the meat counter or wrapped in trays in the refrigerated section at your grocery store. Some common retail cuts include New York steak, flank steak, and short ribs. You can read more about the basics cuts in Chapter 3.

Butchers follow a set of naming standards for the retail cuts they sell. The purpose behind these standards is to ensure that consumers can recognize the cuts they see. That's why renaming cuts or creating new, creative language around meat may be initially enticing, but in the long run, it's not helpful to the industry.

- **Turning the leftovers from the counter into prepared products that can be served at the shop or taken home by customers:** Examples of these goodies include sausage, smoke hams, cured salumi, or in-house specialties like roasted chickens, meatloaf, or ragu. Head to the later section "Getting Familiar with Sausage-Making and Other Preservation Techniques" for an introduction to sausage-making.

- **Setting up the meat case:** The meat case is the "face," or focal point, of a butcher shop. Butchers take setting up the meat case very seriously because it instantly shows customers what they are all about. As such, the meat case is the most important aspect of any butcher shop.

Crafting a beautiful, varied case with delicious-looking meat is a butcher's main source of pride and passion.

If the counter doesn't look attractive, or if the products are not fresh or well displayed, you probably won't keep your customers coming back, and for neighborhood butcher shops, shoppers who return every week are key.

Some butchers may have expertise in all areas of this process. Others specialize in a particular animal or in the further processing of trim and fats. *Further processing* refers to taking semi-finished agricultural products and transforming them into a consumer-ready state. In the meat world, further processing is sausage-making, dry-curing, marinating, smoking, and any of other ways to take meat closer to dinner.

Finding ways to use the entire animal

In addition to being able to effectively cut and further process meat, successful modern-day butchers also need to be knowledgeable about cooking the entire the animal. By using the entire animal through cutting, curing, or cooking, butchers can offset the high cost of carcasses and minimize waste, thus getting the entire animal out the door with the highest profit. To perform these tasks well, butchers must be able to do the following:

- Recognize the most succulent cut to put in the counter
- Know how to use bones to make a rich stock
- Prepare less-valuable meat in a way that makes it delicious (preparing a juicy meatball for a daily soup special, for example)

Helping customers and sharing expertise with others

In addition to all the other stuff he or she does, a butcher is also the person customers rely on for advice and suggestions, helping customers identify cuts, suggesting substitutions, giving cooking advice, and filling special orders.

A butcher may also train and supervise staff, take inventory of the meat, analyze waste and profit, visit the farms he or she works with, place orders for the meat, work the cash register, pay bills, or even sweep and mop the floors at the end of the night.

We need more butchers to train new apprentices, create opportunities for meat education, and breath fresh life back into the trade. If butchers today help pass along their knowledge, we can help preserve the art of butchery. And when more people make the decision to shop with local butchers, more butchers will rise up to meet the demand.

Where have all the butchers gone?

Butchery is a rewarding and honorable trade, rooted in its own heritage and craftsmanship. I have heard many butchers tell tales of how the neighborhood butcher was much like the neighborhood bartender or hairdresser. Most people shopped for food several times a week, so butchers knew their customers well — their stories, their children, their cooking predilections. . . . The butchers may have heard a scandalous story or two, shared thoughtful advice, or donated meat to their neighborhood barbeque. Like Sam on *The Brady Bunch,* they were part of daily life and traditions.

Unfortunately, butchery as a profession has been in decline for a number of years. Until the early 20th century, people who didn't slaughter and butcher their own animals relied on the expertise of the butchers in their communities. All that began to change in the 1920s, when combination grocery stores that sold both perishable and non-perishable items developed. Until then, perishable items were sold in specialty stores such as bakeries (baked goods), butcher shops (meat), dairies (milk), and the greengrocer (produce). Non-perishables, or *dry goods* (flour, baking soda, canned goods, and so on) came from general stores. After the industrial boom of World War II, consumers began to embrace this multi-functional shopping experience. By the 1960s, what I call the "supermarket syndrome" crippled the butcher trade.

Aided and abetted by new meat packing plants that used power saws and mechanical knives and had the capacity to quick-freeze meat in vacuum-sealed bags and large refrigerator trucks able to carry products over interstate highways, cattle was phased off of grass, moved to feedlots, and fed grain-supplemented diets. Other aspects of animal husbandry were lost to industrialization as well, and a new meat industry emerged, one in which meat went from being an agricultural product to a manufacturing product.

The reduced costs and infrastructure made the new system appealing to consumers who could suddenly afford a diet rich in protein. As old-school butchers traded their deep skills and knowledge for retirement, new meat cutters were trained to make the same cut over and over every day. Supermarkets went from cutting *swinging beef* (carcasses hanging on a rail) to ordering *boxed beef* (a box of several of one cut). Butchery went from being work that required expertise and skill to work that required speed and focused repetition. Boxes of broken-down meat were then shipped to supermarkets and merchandized into meat cases. The efficiencies of shipping boxes over shipping unwieldy carcasses changed our entire diet and relationship to the meat we eat.

In recent years, signs point to a another shift in the way we think about the food we eat. Food recalls in the early years of the 21st century and the growing body of evidence linking disease, dangerous pesticides, and foodborne illness outbreaks to unsatisfactory, wide-spread farming practices has gotten people talking about where their food is coming from. The refocused attention on organic and local food movements, as well as the push to embrace sustainable farming, has grown out of the deceptively simple question, "Where does my food come from and how was it grown?"

Patronizing your local shop

Every grocery store, big and small, has a meat counter or a meat case. As a rule, the average supermarket buys boxed meat and then packs popular cuts into trays. If you have read the preceding sections, you know that butchers who run small, local shops do quite a bit more than this. While large supermarket chains can enjoy economy of scale and sell meat for less, purchasing meat from a local butcher shop has several advantages that may outweigh the cost difference. Here is why your local butcher is important:

- ✔ **Neighborhood butchers source meat themselves.** Doing so ensures the meat's quality and supports local agriculture.

- ✔ **Local butchers know that the customer's trust is their biggest ally, and they take your satisfaction seriously.** Selling sub-par protein is not in the best interest of a butcher that wants to stay in business. He wants his customers to be satisfied.

- ✔ **An experienced butcher can introduce you to new cuts and cooking methods.** Most neighborhood butchers are well-versed in the proper preparation of anything they put in their case.

Expand your horizons with new cuts for home-cooked meals. Doing so increases your skills in the kitchen and ultimately leads to eating less processed foods — a step in the right direction if one of your goals is to create healthy diversity and more delicious dishes in your diet.

- ✔ **The butcher handles the meat from beginning to end.** Using all parts of the animal — from aged, bone-in rib eye steaks to sausages to soup bones — yields the most tasty and creative results. This careful use of the carcass is also respectful of the sacrifice each animal makes.

If your local butcher buys and cuts meat from a whole animal ("on the hoof") instead of ordering boxed meat, you know that each cut you see in the meat case came from the same animal. Having cuts from an entire carcass available gives you the option to make dinner match your mood: quick or slow-cooking, fatty or lean, frugal or lavish.

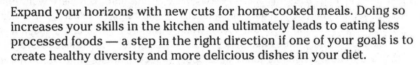

Assuming the Role of Butcher in Your Own Home

Butchering at home is a rewarding effort. With the right knowledge and equipment and a little bit of inspiration, you'll have all you need to take the plunge. And if you make a mistake, don't give up! That's how you learn.

Knowledge and equipment you need

Butchery is a skill that anyone can develop. You just need the right equipment and know-how. Here's the relatively short list of requirements:

- **The proper cutting implements:** You don't need many knives, but you do need ones that are suited for the purpose at hand. If you plan to butcher larger animals (beef or pork, for example), you may also need a saw. I explain all your equipment needs in Chapter 4.

- **A supplier:** Any supermarket or grocery story carries whole chickens for sale, but where do you go when you want a side of beef or a whole pig, or something a little different, like rabbit, which is generally not included in the standard supermarket meat case? You can contact farmers directly to purchase whole animals, or you can contact your local butcher and see if he or she can order a whole carcass for you. Find out more in Chapter 2.

- **Familiarity with anatomy:** To butcher a carcass efficiently and quickly — and in a way that maximizes the amount of meat and minimizes waste — you need a basic understand of the animal's anatomy. Fortunately, you don't need to have a degree in bovine, porcine, avian, or any other type of animal physiology. I tell you what you need to know about each type of animal in the chapters devoted to that kind of meat.

- **An understanding of types of cuts and what you can do with the leftovers:** Despite the unique differences from animal to animal, there are some basic cuts — and terminology — that you need to know. With this general knowledge at hand, your initial forays into butchery will go much more smoothly. Chapter 3 has more.

- **Knife skills:** As a butcher, you need to know how to wield a knife safely. In Chapter 4, I introduce you to the basic cutting techniques and knife safety.

The benefits of butchering your own meat

Beyond the satisfaction of setting out to accomplish a task few people can do, butchering your own meat offers many other rewards. Here are just a few of the benefits, whether you're butchering at home or in a business:

- **Reducing cost:** Buying whole animals is more cost effective if you have the skill to butcher them. Carcasses are sold by the pound at one low cost; waste and labor get added to the cost after processing.

- **Reducing the amount of processed foods in your diet:** Eating fresh meat (and less of it) means that you are not eating as much processed foods. And, if you're using the entire animal (a strategy I recommend), you're doing the processing yourself and are therefore able to better control what you eat.

✔ **Promoting a healthier diet and relationship to food:** We are what we eat, and epidemics related to unhealthy diets and foodborne illnesses have shown that we can't make shortcuts with our diet and expect to circumvent the side effects. Because the protein on your plate reflects each step of the process from "farm to table," you end up with what is truly the sum of all its "parts," healthy, sustainable, or otherwise.

✔ **Getting your children involved in meal preparation:** As you prepare the meat for cooking, your kids can help with the other preparations. Teaching kids how to cook builds self-esteem and independence, instills values, and opens a world of possibilities for future healthy food choices. Learning good food habits in the kitchen is as easy as helping chop the salad greens, setting the table, or stirring the pot. Children feel the weight of their contribution and see instant results: a tasty, nourishing meal to be proud of. Plus, sharing the responsibility for getting ready for mealtime builds a sense of harmony and reverence for the process.

Save meat trim or leftovers throughout the week to make soups or stews. Be creative and let your children help. If family meal preparation is a team effort, the task may not seem so daunting on a busy week night. Also, shopping together with a local butcher may make cooking more exciting.

✔ **Emphasizing mealtime:** By taking the time to prepare a meal, you're sending the message that mealtime is important. Cooking and eating together as a family is important. The act of preparing a meal together is an essential step in building up family roles and traditions and creating healthy future habits.

Studies show that children, especially adolescents, whose families eat together regularly are less likely to be depressed, abuse substances, or have trouble in school. A shared mealtime lets you reinforce your family's cultural values and give your kids your full attention. So try to eat dinner, as a household, three times a week. Then up the number of family mealtimes a week to four.

✔ **Supporting local farms:** When you buy meat in whole carcasses and cut it yourself, you give local farmers the opportunity to flourish.

In order for our meat consumption to flow with nature, we need to give equal value to all the parts of an animal; consider enhancing your knowledge and flexibility by purchasing unfamiliar cuts. Ask a butcher whether he is "long" on anything, and see what culinary adventure ensues!

Rethinking "Go big or get out"

Buying whole local animals from small farms enables the farmer to succeed amidst difficult circumstances. Many small farmers wince knowingly at the expression "Go big or get out." I spoke with Alexis Koefoed of Soul Food Farm in Vacaville, California, who feels that the expression is meant to intimidate small farmers, to make them feel as if something is deficient in their farming methods.

"The perception is that you have to go big, and if you can't get big and keep up, then you need to get out, or you will be squashed. It alludes to the sentiment that small farmers will never benefit from farming 'small.' It is a false premise and a false statement. Small farmers can do good work, and chances are that they will be just fine within the scale that they can manage. Working within the means of their land and labor allows for more small farms to crop up so that everybody can be productive and feed a region or community. The Butcher's Guild and Soul Food Farm both share values of diversity and cooperation to create profits and efficiencies. We believe it is better to go wide than to go big!"

Preserving Traditions: Sausage-Making and Other Preservation Methods

Sausage-making and other preservation techniques originally developed as a way to avoid waste, halt spoilage, and preserve meat to ensure that protein was available during the months to come.

Although we may have access to meat year round, the need to preserve meat is still relevant. With the careful addition of skill, salt, and time, a perishable cut of meat can be cured for future enjoyment. There is also something very soul-satisfying about not wasting good food, and as any expert *charcutier* (an expert in preserving meat through drying, salting, and curing processes) knows, trim from a carcass or excess meat that can't be consumed immediately can be artfully crafted and repurposed into a most delicious, lasting food source.

Making sausage

An ode to sausage: O sweet, O savory sausage, you are a marvelous mixture of meat, fat, and seasoning. Many mouthwatering legacies have been built upon your tasty links.

Whether you feel that sausage deserves its own poetry or not, sausage *is* treasured by cultures world-wide, adored by millions, and known by many names (*saussison, wurst, bangers, franks,* and *wieners,* to name just a few). Traditionally, sausage fillings, made up of the *leavings* (leftovers from the butchery process), were mixed with salt and other spices and stuffed into intestinal casings. Today, sausage may use leavings as well as actual minced cuts of meat. In fact, sausage has become so popular that many butchers set aside meat just for this purpose. In addition, butchers may stuff modern sausage into either intestinal or synthetic casings. The wide range of sausage styles available to consumers today draws on cultures from all around the world.

Here are some different kinds of sausages:

- ✔ **Cooked or raw (fresh) sausage:** Cooked sausages like hot dogs have a longer shelf life and are faster to grill or sear because they are pre-cooked and need only be reheated before consuming. Raw or fresh sausage, like mild Italian or breakfast sausage, have a shorter shelf life and must be cooked completely through before eating.

- ✔ **Forcemeats:** Forcemeats are a mixture of meat, fat, and seasoning that are emulsified together, usually through grinding or pureeing. Examples of forcemeats include pate, terrine, and roulades.

- ✔ **Emulsified sausages:** Emulsified sausages are cooked sausages whose fillings have been processed into a fine paste. Examples of emulsified sausages include hot dogs and bologna.

- ✔ **Dry-cured sausages:** Dry-cured sausages can be both sliceable and spreadable. Examples include salami, Spanish chorizo, and summer sausage.

For modern butchers, sausage-making is now more art than task, and a butcher who makes sausage is often a specialist. Sausage-making is where butchers' creativity, skill, and craft can shine. Although a butcher may have some clever time-saving tricks or fancy knife-wielding skills up her sleeve, only so much creativity can be applied to cutting raw meat. By developing house-made sausages, marinades, and specialty cured or value-added products, butchers can create something from raw meat that is exclusively theirs. You can find instructions on sausage-making techniques and several sausage recipes in Chapters 14 through 16.

In the 15th century, local European guilds worked with food craftsmen to regulate the production in each of their industries. While butchers' guilds regulated the trade of meat in one region, members of the charcutiers' guild used preservation methods to create traditional dried, salted, and cured meats, each with their own regional distinctions.

Other preservation techniques

Several other food-preservation techniques continue to be used today. The age-old techniques of crafting meat, fats, seasonings and a variety of other ingredients such as nuts or fresh produce into baked forcemeats are still relevant to modern day preservation. The general difference between sausage-making and these other techniques is that these items are baked or poached in dishes instead of casings. Some examples of foods created through these processes include the following:

- **Pâté:** A pâté is essentially a meat paste, commonly made of liver or poultry (foie gras or chicken liver, for example), but the term *pâté* more accurately means pie, or savory pie.

- **Terrines:** A terrine is a dish made up of a mixture of meat, fish, or vegetables that's been cooked and allowed to cool and set in its container, which is also called a terrine.

- **Galantines:** Galantines are dishes in which a combination of white meat or fish, vegetables, and nuts are arranged in layers, cooked, pressed, and served in *aspic*, a kind of gelatinous broth.

- **Roulades:** Think of meat pinwheels: Typically in a roulade, a soft filling is spread over a flat piece of meat or fish, which is then rolled up, creating a spiral when sliced.

- **Salt-cured and brine-cured products:** In these products, meat or fish is preserved in a salt brine and then smoked or roasted. Examples include country hams and corned beef briskets.

You can read more about these preservation techniques in Chapter 17.

Promoting Healthy Food Systems

Although as consumers, we may feel powerless to control the trends, prices, availability, or selection of products, in fact, "voting with our forks" is a very effective way to promote change. Consumer dollars do the talking and have been the impetus for great change in the supermarket. By consistently purchasing organic or local products, you create a demand that your grocer and butcher will fill by buying more of those products. If more butchers buy meat raised organically, for example, more ranchers will produce that kind of meat. And slowly, the food system starts to change. Making deliberate choices about what you eat gives you a greater degree of control over what foodstuffs are made available.

The Butcher's Guild: Building community and promoting healthy food system

In 2010, I met with Marissa Guggiana, fourth generation meat purveyor, cookbook author, food activist, and now my business partner and co-founder of The Butcher's Guild. Marissa and I met to discuss developing a food community focused around butchers, with the purpose of promoting the future of healthy local food systems.

Marissa had just finished writing her book, *Primal Cuts: Cooking With America's Best Butchers,* during which time she had traveled the country and chronicled the stories of this new generation. As she and I talked, we identified a common focus among the butchers Marissa had encountered and our own peers: a renewed passion and excitement for the trade, a hunger for education, and a deep personal investment in the preservation of butchery. We quickly realized that this group was aligned and yearning for community.

So we created that community: The Butcher's Guild.

We chose to include a slightly broader spectrum of meat industry professionals in our definition of a *butcher,* the criteria being simply that they butcher whole animals in their businesses. Our reasoning? It takes chefs, charcutiers, caterers, and butchers to find the most profitable and delicious home for a whole animal, and anyone who uses a whole animal is butchering ; there's no way around that.

Our members hail from all over the country. They are innovators, old-school masters, budding talents, and eager apprentices. We help each other with innovations, advice, and a shared vision of healthy, sustainable food systems.

Our motto? Fraternity – Integrity – Community. That pretty much says it all.

Food is a personal choice, and it's not my business to presume to know what is right for anyone other than myself. But as the co-owner of my neighborhood butcher shop, I have seen good food in action, have had the good fortune to watch the community grow, and have felt the monumental rewards of being a part of it. By butchering your own meat, you can, too.

Chapter 2

Meat Is Meat, Right? Wrong!

- -

In This Chapter

▶ Discovering basic cuts and the meaning of common meat labels

▶ Knowing the factors that affect flavor

▶ Going beyond flavor: sustainability, humane treatment of animals, and more as quality markers

- -

Meat tastes good, but is all meat *good*? When you take the first bite into a superb steak or a succulent sausage, you can instantly judge quality. But quality runs deeper than your taste buds. The way the animal was raised and the way the carcass has been treated affect the taste, the nutritional value, and the possibility of any adverse affects to your health.

This chapter gives you all the information you need to make educated decisions at the meat counter. Here, I tell you how to determine what elements make meat good, how to decipher meat labels and the mysterious nomenclature of retail cut names, how to understand what affects the flavor of meat, and how fat content, marbling, and aging can lead to delicious, decadent results.

Knowing What You're Getting

Rib eye, standing rib roast, export rib, bone-in rib chop, boneless rib chop, prime rib, cowboy steaks . . . all these cuts are either bone-in or boneless variations of the same section of muscles from the forequarter of beef. But if you don't know the difference between one cut and another, or if you're not aware of key similarities, how can you make a good buying decision or a decent substitution when the cut you're looking for isn't available?

Obviously, the most reliable way to get the best bang for your buck behind the meat case is to know what you're buying and how to best prepare it. With this info, feeding a family on a budget is much easier because you'll be able to confidently choose less-expensive, quality cuts and substitute meats without sacrificing taste or tenderness, and you'll be empowered to embrace a little off-the-cuff culinary creativity.

You say "tomato"; I say "porcupine" — Playing the name game

The meat beat isn't changing much because not much new and exciting is happening when it comes to body structure. Unless cows start spontaneously growing extra body parts, meat is what it is. Still, the temptation to give fanciful names to established retail cuts is alluring because it gives consumers the appearance of a new, exciting, or special meat cut. But the multitude of over-lapping names and terminology around meat also creates confusion for shoppers. Calling a New York strip a "Harrison steak," for example, is hardly helpful and makes knowing what's what really hard.

Having a standardized language around meat is important. Butchers and meat processors rely on NAMP (North American Meat Processors Association) to create common industry definitions for retail cuts. The NAMP *Meat Buyers Guide* is a book of recognized names, cut specifications, and identifications for the U.S. meat market. Other countries have their own versions of these standards. Although the NAMP guide may be a very useful resource for butchers, for most home cooks, it feels like an overwhelming amount of information and technical terminology that is irrelevant to the shopping experience.

When you shop for meat, shop with knowledgeable butchers you can trust. Doing so ensures that, when you have a question about how to prepare or substitute meat cuts, you get the information you need. Reading this book and learning about what cuts are and where they are located also give you an advantage.

Think cooking instead of cutting

Meat diagrams are useful visual representations of how carcasses are broken down into retail cuts, but not all cuts are included. Familiarity with primals, subprimals, and retail cuts are important, but researching and memorizing each of the cuts (and their variations) can be a daunting task. In addition, every region and butcher counter puts its own spin on nomenclature and selection.

To simplify things for consumers and novice butchers, try this: Look at an animal in simple cooking preparation terms. The idea is that everything on an animal can fall into these three preparation categories: grilling, braising, or roasting. All are universal culinary techniques and apply to all cuts. Looking at meat this way is a good start to understanding meat and the different cuts. This perspective can help you interchange cuts and choose the best method of preparation for your mealtime recipes.

How names become standardized

Kari Underly, of Range, Inc. in Chicago, is a third-generation butcher, an expert on cut nomenclature and meat musculature, and the author of *The Art of Beef Cutting*. She shares her insight and experiences in one of the study groups responsible for identifying and naming the Boneless Beef Chuck, Under Blade, Center Steak, a.k.a Denver Cut.

"Porterhouse, filet mignon, flat iron. . . these names are so familiar to us now we often forget where they came from. Yet, 'How do cuts get their names?' is one of the most popular questions I get asked. The easiest way to explain the answer is to walk you through an example of the process — from carcass to case.

"**The task:** While consulting on a Beef Checkoff project, I was responsible for finding new value in underutilized subprimals. What I found was the Chuck Roll. Through detailed muscle profiling, we discovered it was the fourth most tender muscle in the beef animal. With some creative blade work, I fabricated a lightly marbled, rich-tasting steak.

"**The name:** This attractive new cut needed both a fanciful name for consumer use and an anatomical name so butchers, chefs, meat execs, restaurateurs, and the like understood where this cut came from and could purchase it. Sounds simple right? Believe me, this is an arduous process. Just generating possible names requires an understanding of the cut's attributes and potential uses, knowledge of current cuts and beef eating trends, insights into consumer preferences, and an understanding of Uniform Retail Meat Identity Standards (URMIS). All current cut names and UPCs and/or PLU codes are located in the URMIS program, and new cuts must be presented and accepted before becoming 'official' products.

"After research, focus group studies, and presentation to (and approval from) the URMIS board, I had a winner:

"**Fanciful name:** The Denver Cut

"**URMIS name:** Beef Chuck, Under Blade, Center Steak (Denver Cut), Boneless

"**UPC and/or PLU code:** 1098 and/or 1913

"All of this information is essential. A new product must have a unique name and description in order for sales and mark-down data to be properly analyzed. With this information, meat executives and butchers can analyze market-level data and compare it to other retailers in the area."

If you're not sure where on an animal a cut comes from, ask the butcher whether it should be grilled, braised, or roasted:

- ✔ Tender cuts are often grilled (or seared).
- ✔ Cuts of medium texture can be roasted and sliced thin to maximize the tenderness of the cut.
- ✔ Tough cuts should be braised or stewed; these slow-cooking methods break down and soften the fibrous muscles, tough connective tissues, and tendons in the meat.

A working muscle is a tougher muscle, and if you can envision animals in movement, you can figure out which muscles work and which muscles don't work (as hard, anyway). Because of the increase of blood flow to these working muscles, they are richer in collagen and have more depth of flavor.

Now suppose you walk into a butcher shop armed with your favorite fajita recipe only to discover that the butcher is fresh out of the three pounds of flank steak you need. What do you do? Ask for a quick-grilling meat: Think cooking instead of cutting!

Take a moment to study a well-stocked meat case. You'll see that some cuts are sliced thin; others are cubed, ground, or left intact as whole muscles; and roasts may be both deboned and left bone-in. Ask yourself why the butcher cut the meat in a particular way. A butcher's expertise is isolating muscle groups and cutting the meat into portions by considering how they will be best eaten. The next time you look at the trays of meat in a retail case, keep these basic rules in mind:

- Grilling, searing, or quick-cooking cuts are typically sliced thin or cut into individual steaks.

- Roasting or braising cuts are usually larger, left in whole-muscle pieces, or cubed into stew meat.

Deciphering labels

A label makes quick work of telling product basics: What is it? How much does it cost? Where is it from? And so on. The purpose of a label (after catching the customer's eye) is to provide relevant information about the product that helps customers make a decision about whether they want to purchase it.

In regards to fresh meat, labels tell how the meat was produced and give indications of how the animal was treated, what it was fed, and how it tastes. Labels can be confusing, though, because they can be vague and sometimes misleading. In addition, some labels are certified by the awarding organization and follow their own set of standards and procedures; others are unaudited, with no third-party certification, and are not recognized by the United States Department of Agriculture (USDA) or Food and Drug Administration (FDA). Following is a list of labels you may see at the meat counter. Use this knowledge to find the products that work for you:

- **Grass-fed:** This label means that the animals were raised on grass and hay, and it applies to ruminants (cattle, sheep, goats, and game). It does *not* mean that the feed was organic. One common misconception about grass-fed meat is that the animals are raised 100 percent on grass, when, in fact, this may not be true. Cattle can graze for most of their life and then be finished on grain for several weeks.

✔ **Grain-finished:** This label indicates that the animal was raised on grass but then *finished*, or fattened, with grain.

✔ **Pastured:** This label applies only to poultry and pork and simply means that the animal was allowed to forage for food outside. It's not related to organic standards.

✔ **Vegetarian-fed:** This label means that the animal had dietary access to only vegetable proteins and was not fed animal proteins.

✔ **Free-range:** This label applies to poultry and simply means that the poultry has access to the outside. Be aware, though, that this access could mean an open door at the end of a barn; it doesn't indicate that the poultry has spent any particular amount of time outdoors.

One popular misconception is that the free-range label implies a foraged diet. That is not necessarily the case because the label has nothing to do with the poultry having been grass-fed. A free-range bird wasn't in a cage, but it may have spent its entire life in a barn.

✔ **Halal and Zabiah Halah:** To be labeled "Halal" or "Zabiah Halal," the products must be prepared in federally inspected meat packing plants and require handling according to Islamic law and under Islamic authority. In addition, "Halal" animals must be slaughtered without being pre-stunned.

✔ **Humane treatment:** This label means the animal was raised according to the standards mandated by the Certified Humane organization which "allow the animals to express their natural behaviors. Chickens can dustbathe, pigs can root, cows can graze, and all farm animals are able to live their lives with the space, shelter, nutrition and the care they need."

✔ **Animal Welfare Approved:** AWA is a national nonprofit organization that audits, certifies, and supports farmers who raise their animals according to the highest welfare standards. It's the most highly regarded food label relating to animal welfare, pasture-based farming, and sustainability.

✔ **Kosher:** This label indicates that the meat and poultry products were "prepared under rabbinical supervision." Animals must be slaughtered without being pre-stunned.

✔ **Organic:** This label indicates that the animals were fed only organic grains. To use this label, producers must meet federal organic standards, which, at this time, mean that the animal may have been raised in a facility similar to a factory farm but fed only organic feed.

- ✔ **Natural:** This label doesn't have anything to do with the animals' diet. It means that the product cannot contain any artificial or synthetic ingredients, that no dyes were added, and that the meat was "minimally processed" (translation: the product wasn't substantively altered during processing).

- ✔ **No hormones:** To label a product "hormone-free," producers are required to provide sufficient documentation that no hormones were used in the growth or life of the animal. Hormones are not allowed in raising hogs or poultry.

- ✔ **No antibiotics:** For red meat and poultry, the term "no antibiotics added" is allowed if the animals were raised without antibiotics. To substantiate this claim, the producer must submit compliance records to the Food Safety and Inspection Service (FSIS) division of the USDA.

Do you prefer grass- or grain-fed meat? Organic or local? How you answer is really a matter of personal taste and conviction. The American palate, for example, is accustomed to grain-fed meat. Grass fed is more natural but leaner and cooks faster, and some people don't like the flavor. Depending on your own personal preferences and convictions, you need to understand labels and decide where to invest your dollars.

Focusing on Flavor

Although fat, especially intramuscular fat, or *marbling,* can definitely make meat juicier, richer, and more texturally appealing, it is not solely responsible for flavor. Breed, diet, aging, the life of the animal, and an invisible fat called phospholipids also contribute mightily to the flavor and texture.

In this section, I explain what elements affect the flavor of meat and how meat is graded to predict flavor.

The amount and kind of fat

Everyone knows that the tastiest meat is marbled, right? Those thick veins of fatty goodness translate into tenderness, taste, and rich juiciness. Well, that's true in general, but what you may not know is that meat also has an invisible fat that's an important component of a meat's flavor and texture.

Meat science

The term "meat science" probably conjures images of pork chops in petri dishes, scientists in white coats, T-bones hooked up to metal probes while lasers shave off paper-thin slices, and slabs of big, beefy muscle wired to computers that analyze for maximum deliciousness. These images may sound like surreal fantasies, but they are not actually that far from the truth.

Meat scientists analyze meat. Simply put, meat science is the combination of animal science, food science, and muscle biology as related to the production of muscle food products. Meat scientists also may specialize in animal husbandry and production, meat manufacturing and processing, microbiology, Hazard Analysis Critical Control Point (HACCP) mandates, and

more. Regardless of the area of focus, all meat scientists focus on the systematic knowledge, observation, and research surrounding meat in order to figure out how to produce "better" muscle and value-added products in a more efficient and consistent manner.

Note: HACCP plans document steps and safety procedures designed to ensure that meat is handled properly and food safety guidelines are followed during the processing of meat. All businesses involved in the wholesale meat trade are required to implement HACCP standards and keep daily records of their compliance. This mandate ensures the food safety and security of the meat-eating public.

Marbling, the fat you can see

The first flavor factor is fat, or the lack thereof. *Marbling* is intramuscular fat dispersed within meat. "Fat is flavor" is a universal meat chant. After all, well-marbled center cuts are the moist meat gems responsible for the term "big, juicy steak." Braising or stewing cuts may be either lean or fatty, but fat in a braise yields a more succulent, fork-tender end result. Fattier braising cuts like brisket or pork butt are known as the reigning champions of slow cookery for their decadent fat-to-meat ratio.

Meat's fat content differs from animal to animal, and in each animal, fat content changes from part to part. Muscles that are used often consume stored-up fat, meaning the meat from the hardest working areas don't have much fat. Parts of the animal that aren't used as much don't spend energy, so they produce fattier cuts.

Because fat is a key component of many of our favorite meat dishes, folks tend to think of lean meats as dry and flavorless. The truth is that lean cuts can be just as yummy as more marbled cuts. The key is the preparation. Leaner cuts, because they're more dry than fattier cuts, benefit from brining, marinating, or tenderizing.

Phospholipids — the fat you can't see

Phospholipids store energy within muscles and are a primary component in cell membranes. Although they're not visible to our eye, they pack more flavor than visible marbling. During the 1980s, a scientific study was conducted to surmise what exactly fat had to do with flavor. Researchers found that when visible fat was removed from the meat, the flavor was unaffected, but when the phospholipids were extracted from the meat, a very different — and unpalatable — protein remained.

The age of the animal

In addition to fats, seen and unseen, age is a big determination of flavor. Organic materials, like *terpenes* (organic compounds found in many plants), are absorbed into the animal's flesh as it ages. Terpenes are often strong smelling, and as the animal grows older, these organic materials increase the gamey quality of the meat.

The meat's grade

Across the globe, organizations like the USDA or the European Food Safety Authority (EFSA) regulate food safety and quality. In the United Sates, the USDA grades beef by running a complex calculation that takes into account several components, two of which are the amount and distribution of marbling in the rib eye section and the age of the cow.

Grading is a lot more complicated than looking just at the marbling and age. Other things the USDA takes into consideration include percentage of fat to meat; fat thickness; kidney, pelvic, and heart fat; carcass weight; maturity of the skeleton; appearance and color of the ribs; condition of the chine (spine) bones; color and texture of lean meat; and carcass yield.

Quality grading is a prediction of palatability (flavor, juiciness, and tenderness). Flavor and juiciness are predicted from the marbling in the rib eye, and tenderness is predicted from the age of the animal at the time of slaughter. The USDA has eight quality grades for meat, in order from most to least desirable:

- Prime
- Choice
- Select
- Standard

- ✔ Commercial
- ✔ Utility
- ✔ Cutter
- ✔ Canner

Bottom line: Of all the qualities that contribute to meat magic, grading comes down to the fat content of one muscle. So let grading be one arrow in your decision quiver, not your entire arsenal.

Not all animals are graded. Pigs, for example, aren't quality graded. Lambs are sometimes graded, and the vast majority of them receive a Choice rating. Chicken, rabbit, and goat are also not graded.

Whether the meat is dry or wet aged

All meat must be aged. Aging, also known as *conditioning* the meat, enhances the flavor and improves tenderness. Aging meat is the process by which microbes and enzymes work through the muscles to break down fibrous connective tissue. The length of aging varies from animal to animal and can come down to economic, as well as flavor, considerations. In beef, aging is achieved via two standard aging methods: dry and wet aging. Some people prefer the stronger flavor that results from dry aging, whereas others like the less intense flavor that comes from wet aging.

Dry aging

Years ago, dry aging was the standard. If meat was not cured or canned, it was dry aged. To dry age meat, you hang it up to dry for 21 to 28 days. During the aging process, meat loses a lot of moisture, and the loss of moisture concentrates the flavors, leaving the meat with a deeper, funkier, richer taste. Dry aging also helps to tenderize the meat. While the meat hangs during aging, an inedible outer crust forms on the surface of the meat, and enzymes in the meat break down the tissue, making the meat more tender. The crust that forms does not spoil the meat, and it's always cut away before the meat is sold.

Dry aging increases the cost of the meat because the producer incurs the cost of trim loss and water loss (weight loss) as the meat dries out. Dry aging also means that the producer ends up paying for real estate (for storage) and has to wait much longer to see a financial return on the product.

Dry-aged beef may not be as common as it once was, but you can still find it at steak houses and some high-end butcher shops.

Wet aging

These days most meat that you find in the supermarket has been wet aged. Meat that is wet aged is put into a vacuum-sealed bag and ages usually during transport. This aging method prevents the meat from losing moisture, but it also keeps the flavors from being concentrated. The process of wet aging takes only a few days.

Wet aging is more popular because it's faster and because the meat doesn't lose weight due to loss of moisture. Wet aging also results in a more mild flavor than dry aging. In addition to a less-intense meat flavor, the flavor of the blood is much more prominent in wet aging.

Ensuring you get the best flavor

As I note in the preceding sections, flavor is impacted by a number of factors. Therefore, when shopping for meat with great flavor, follow these guidelines:

- ✔ Look for meat that comes from a source you can trust and ask about the diet of the animal (grass-fed meat is leaner). An animal's diet affects the flavor of the resulting meat because substances are concentrated in an animal's tissues.

- ✔ Steer clear of factory farmed meats, which are fed a cheap diet. Unsavory living conditions also means that these animals can accidentally end up eating things they shouldn't (like feces or other animals). Instead, try heritage breeds (turkeys, pigs, and chicken) or game meats, which are typically farmed on a smaller scale with more attention to quality of feed and farming environment.

- ✔ Ask your butcher how long the meat is aged. Get to know your preference on the fresh-to-aged continuum.

- ✔ Get recommendations from your butcher. A question as simple as "How is the flavor of the meat?" can yield the info you're looking for.

Of course, how you prepare the meat also impacts its flavor. Gregg Rentfrow, butcher and meat science professor at the University of Kentucky, believes that Americans tend to over-season their meat. He offers this advice on getting the right amount of seasoning to enhance and not overshadow the meat's natural flavor: "Well-marbled meat needs limited seasoning, with a very basic spice mix (salt, pepper, and garlic powder). In this case, the purpose of the basic seasoning is to enhance the natural flavors. However, lean meats are great for a marinade, due to their ability to carry flavor. Because lean meats don't have the background flavors from fat, the seasonings and/or marinades play a larger role."

Broadening Your Definition of "Good"

Good meat should be delicious, but you need to consider more in matters of meat than flavor alone. Now's the time to broaden your perception of what makes food good.

"Good" is quantifiable. In matters of meat, these standards should be the humane treatment of livestock, adherence to sustainable farming practices, practicing respectful and skillful butchery, and the appropriate cooking or curing of meat.

For agriculture to be sustainable, it must meet the following standards:

- ✔ Be nourishing and meet human food and fiber requirements

- ✔ Efficiently use non-renewable resources and on-farm resources, integrating appropriate natural biological cycles and controls

- ✔ Augment environmental quality and natural resources based upon that which agricultural economics depend

- ✔ Be profitable and sustain the future of the farm and its operations

- ✔ Increase quality of life for farmers and society as a whole

Ask your butcher what farms she buys from. What you'll probably find is that she works directly with farmers to purchase her meat and makes on-site visits to the farms and ranches. She should be able to give you more information about whether or not a farm is sustainable.

The Good Food Awards

It is becoming increasingly easier for us, as consumers, to identify the qualities of food, as we re-define (and label) our standards for food in general. Sarah Weiner, director of the Good Food Awards, is a leader in labeling good food. Her organization's mission is to promote, identify, and reward good food and its practices. The Good Food Awards honor craft producers whose products not only taste great but also adhere to good business standards identified by the organization. Here's what Sarah has to say about what Good Food stands for:

"Good Food is about taste, and much more. It's about cultural traditions being kept vibrantly alive through crafting and enjoying country ham and chorizo sausage. It's about people all along the supply chain, from farm hands to butchers, being treated fairly and taking pride in bringing something nourishing to their community. It is the sort of food we all want to eat: tasty, authentic, and responsible. Of the dozens of choices we make on a trip to the grocery store, it is at the butcher's counter where one decision has the most far-reaching impact on your own health, the planet, and the community."

TECHNICAL STUFF

Love meat but eat less of it!

On average, we should all be eating less of this good meat every day. The average American eats about 200 pounds of meat a year, the average Canadian eats 137 pounds (62 kilos) per year, and the average Australian eats 242 pounds (110 kilos) per year. Yet experts suggest that to get the protein we need we should eat 4 to 5 ounces of meat four times a week, which works out to be a little over a pound a week; or 52 pounds a year.

Eating less meat happens naturally when you purchase meat from local butchers because, ultimately, small-scale farming produces more expensive meat and forces you to buy smaller amounts in order to stay within your budget. And buying smaller amounts of meat helps you balance your diet.

Although shopping with your local butcher may help you slim down and promote better meat eating habits, it's also good for the animal, good for the farmer, good for the consumer, and good for your community. A chain reaction occurs as more and more people begin to support local butchers. More demand and consumer scrutiny breeds more of the right kind of supply. Chances are, if you shop with a butcher who breaks down whole animals in his business, the butcher knows the farm, knows the meat he's selling, and takes pride in the skillful preparation of that meat. These extra considerations assist you in knowing how to identify good meat. It just tastes better when it's all good.

Chapter 3

Cuts and Terminology: The Basics of Butchery

*I*n the past, butchery was the act of preparing meat for sale and included slaughtering the animal; breaking it down into primal cuts, subprimal cuts, and retail cuts; and performing further processing, like salt curing, smoking, and creating country pâtés.

Nowadays butchers don't necessarily work on the animal from slaughter to retail case. Meat may come to the shops already broken into primal or even subprimal cuts. In most cases, butchers never deal with an entire carcass but instead buy meat in sides or quarters. If you are butchering meat at home, chances are it's chicken or larger cuts of pork or beef.

Still, traditional butchery is going through a quiet resurgence, in which the modern butchers are creating relationships with small farmers, with slaughterhouses, and with their customers in order to offer customers better meat and better information.

Home butchers — whether you plan to butcher the livestock you raise yourself or source your meat and poultry from a local farmer for butchery — are a key part of the return to the kind of meat production that takes responsibility for the whole process. In this chapter, I explain a few basics about cuts and butchery terminology, information that will help you as you begin your butchery endeavors.

Breaking It Down the Easy Way: Meat Maps

Before you even pick up a knife, you need to understand what you are looking at. Animal diagrams, or *meat maps,* are visual representations of the carcass that show the primals (and sometimes the subprimals or even retail cuts) on the animal. The purpose of these diagrams is to give consumers a point of reference for their purchases. Figures 3-1 through 3-4 show the typical cuts for chicken, lamb, pork, and beef. (You can find even more detailed information regarding primal, subprimals, and retail cuts common to each type of animal in each of the respective meat chapters.)

Different countries have different diagrams that are unique to their food cultures and butchery practices. How an animal is butchered depends entirely on how it will be cooked. In the United States, the typical mealtime involves small, nuclear families and quick turnaround dinners. For this reason, in the U.S., small, grilling cuts are favored over large roasts. Additionally, not all countries value all cuts the same. In Brazil, the cullote steak (top sirloin cap) is highly prized and valued; in the U.S., this cut is not as highly prized.

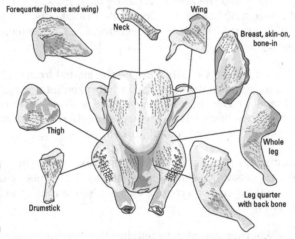

Figure 3-1:
Chicken
cuts.

Illustration by Wiley, Composition Services Graphics

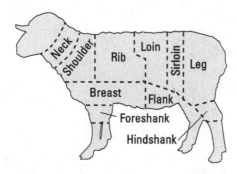

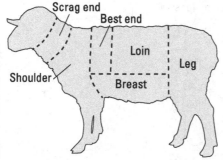

Figure 3-2:
Lamb cuts
in the U.S.
(top) and
in the U.K.
(bottom).

Illustration by Wiley, Composition Services Graphics

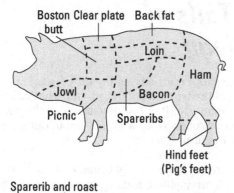

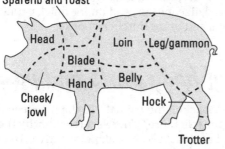

Figure 3-3:
Pork cuts
in the U.S.
(top) and
in the U.K.
(bottom).

Illustration by Wiley, Composition Services Graphics

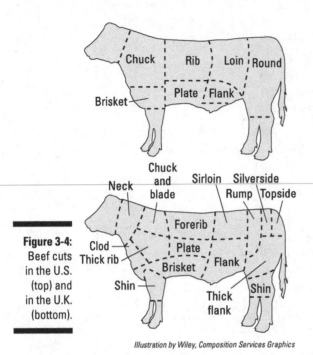

Figure 3-4:
Beef cuts
in the U.S.
(top) and
in the U.K.
(bottom).

Illustration by Wiley, Composition Services Graphics

Making Heads or Tails of Butchery Terminology

Gather your tools, your skills, your curiosity, and your courage, and dive in. Butchery is an integral part of the way we eat. No matter what you believe in, believe that without the hands, heritage, and hearts of crafts people, we are lesser for the journey. Let's repair what we have lost in our supermarket stupor and work harder for the future of our food.

When I set out to write this book, I immediately thought about breaking down the barriers of meat nomenclature, which, frankly, can be very confusing, especially for a non-book-nerd, like me. So read on for quick definitions of terms you'll encounter in this and other butchery books.

Keeping track of body parts and positions

Why not just say the front or back of the animal? Or the big muscle here or the little triangular shaped muscle there? There's a reason why the meat industry keeps the names of each cut so clinical: perspective. To clearly articulate what's being described and where it is, many butchery books rely on clinical terms, like *anterior* or *posterior*, and the names of bones or muscles.

The Butcher's Guild philosophy

Before I get into the practical basics of butchery, I want to share the most fundamental aspect of the process for me: the philosophy behind what I do.

To be a good butcher, you must have a good heart, a good source, a good hand, and a good voice:

- **A good heart** means conducting your business with integrity. Be honest, be knowledgeable, and be respectful to your customer and to yourself. Medieval butcher guilds were based upon a moral code of behavior to fellow butchers, as well as a standard of sale. In this age-old business practice, relationships, butcher to butcher and butcher to customer, are built on trust, trust that will keep customers coming back time and again. Sadly, the relationship of trusted butcher to consumer was nearly snuffed out by the supermarket era's reliance on self-service meat cases. Shoppers had to rely on labeling or brands to assure them of the quality of their meat. When trusted butchers stopped standing behind the counter and handing you your steak with a smile, the connection grew as cold as pre-packaged steak in the meat aisle. But every time a butcher sells you an honest-to-goodness quality cut of meat with a cooking suggestion and a bad joke, an angel gets its wings.

- **A good source** means selling healthy, responsibly and humanely raised, nourishing meat. Small businesses cannot compete with large supermarket chains in efficiency of production or buying power. What they can offer instead is better quality products and experiences. Part of the experience is the story. Tell your customers about the farms, tell them about meeting the farmer at 4 a.m. to pick up the pig, tell them about the breeds and about the smell of the hay at the end of summer. You can't shrink-wrap that.

- **A good hand** refers to the skill it takes to butcher well and responsibly, and the commitment to continue sharing, learning, educating, and striving for increased knowledge, less waste, and more creativity. A butcher's cut-work is the focal point of his passion. Butchering is a Zen-like process in which the butcher can quietly focus on the singular task at hand. I have watched many butchers cut, each in their unique way, and have enormous respect for this diversity. Although there is no one "right" way to butcher, there *is* a right way to conduct yourself as a butcher: Work hard, respect the life of the animal, hone your skill, and teach what you know.

- **A good voice** is one that speaks to and for its community. Investing in your community is a transaction that will come back to you tenfold, and it is a central concern to the Butcher's Guild. If you are an independent butcher or small business owner, get involved in your community, barbecue some sausage for your local elementary school fundraiser, donate meat to the retirement home, give the biggest, best turkey you have to a low-income neighborhood family, participate in your local food community — and see where the handshakes and open ears take you. We all have many communities: our businesses, our towns, our families, our industries. No matter what the group is, you *have* to give to get.

These tenets and ideals are what keep me passionate and focused every day and why I wanted to write this book. The rewards of these tenets are all-encompassing, whether you are an experienced butcher or just starting out. In the Butcher's Guild, we believe that, when these four tenets are respected and developed, they foster the individual, the business, the industry, and the integrity of the entire food system. In fact, to be inducted into the Butcher's Guild, all candidates must sign an oath swearing to uphold these tenets. Keeping these principles in mind will set you on the right path to meat mastery.

If you're a home butcher, you don't necessarily need to know the anatomical names for the body parts, nor do you need to know these terms to successfully follow the instructions in this book. But here are some things you'll want to keep in mind about butchery terminology:

- ✔ **The common use of "anterior" and "posterior":** These terms tell you where on an animal a particular thing is. *Anterior* means "the front" or "near the front," and *posterior* means "the back" or "near the back."

- ✔ **The use of common, traditional, and anatomical names for the parts:** In this book, I call a leg a leg. Feet are known as *trotters*. In other butchery books, scientific or anatomical names are often used. For example, the anatomical names for the piece of meat you know as a beef tenderloin or filet mignon is *psoas major* and *psoas minor*.

If you're a professional butcher (or plan to become one), you should bone up on the proper terms and how to read acceptable ranges of error. For example, "Make a cut no more or no less than 1 inch from the bottom of the scapula (or parallel to the spine, and so on)." These descriptions are allowances of error that create the space bordering a subprimal or retail cut so that you may mentally map out how it should be produced. Although this information may be difficult to digest, don't fight it. Become one with the meat geeks.

Understanding cut terminology

Butchers throw around a handful of terms while "cutting it up" at the butcher block, and knowing these terms will be helpful to you. Although the following definitions are just a few of many, they help you understand butcher lingo and get you "in" with the meat crowd:

- ✔ **Breaking:** Butchering meat down from a whole side or quarter

- ✔ **Side:** Half of a carcass

- ✔ **Quarter:** Quarter of a carcass

- ✔ **Primal:** The first, or primary, portions that the carcass is broken into

- ✔ **Subprimal:** Cuts larger than steaks or roasts; typically isolated muscles

- ✔ **Retail cut:** Standardized meat cuts you find in a retail meat case

Breaking news: Bench (or table) or hanging

You can butcher an animal while it's lying on a table or hanging from a hook. Which method you choose depends on the kind of animal, its size, and your own comfort and skill level. The next sections have the details.

Breaking on the table or bench

When you butcher an animal on the table or bench, you cut the carcass (or, for example, the forequarter of beef) as it lies on a large cutting table.

You can butcher any animal on a bench, but bigger animals are better broken on the rail. In this book, most of the butchery you do (poultry, pork, lamb, and rabbit) occurs on the bench due to the size of the animal.

Breaking a hanging carcass

When you butcher a hanging carcass, the carcass (or side or quarter) is attached to a rail or A-frame and hangs from a hook. You then cut it into pieces as it hangs. This method is typically reserved for butchering large, heavy animals because it's much more ergonomically efficient. Gravity assists you when you are removing large portions of the carcass that are heavy and cumbersome. In this book, you butcher beef fore- and hindquarters this way. (With smaller animals, whether you butcher on the bench or rail may just be a matter of preference. In my butcher shop, for example, we butcher lamb and pigs hanging from the rail, too.)

One of the key things to remember about butchering a hanging carcass is to be prepared to catch the weight of the piece you've just cut free. Have your boning hook safely anchored between a rib or thick section of the piece you are removing. For detailed instructions for breaking a hanging carcass, head to Chapter 12.

Sourcing the Freshest Cuts from the Supplier or Meat Counter

The freshest cuts of meat are the ones you cut yourself, and if you can't cut them yourself, the next best thing is to know the person who did. And after reading all the information in this book, you'll know how to shop for meat and what questions to ask the butcher; you'll know how to read meat labels and be flexible with your purchases. Make no mistake: You are a discerning customer deserving of the best-quality meat!

Finding a reputable supplier

So now that you are ready to cut up some meat, where do you find it? Well, the answer is quite simple: Find a farmer in your area and contact him or her directly. If you can't find a farmer to work with, call around to all the small, local butcher shops in your area and see whether they can order a whole hog or lamb for you (or direct you to a farm or supplier that can). Whole animals are sold at one flat price per pound.

Make sure to ask questions about the farm and its practices. Many butchers take field trips to the farms they do business with to make sure the product they are buying is up to snuff. Some butchers even refuse to sell the meat if they haven't visited the farm themselves.

Judging freshness at the meat counter

Shopping for good meat can be challenging. Never mind the fact that retail cut names are confusing and often redundant (refer to Chapter 2 for info on how retail cuts are labeled). People tend to cook less nowadays and are therefore less knowledgeable about how to prepare the amazing array of cuts available. Lack of exposure to different cuts and the overabundance of pre-packaged meat has also made it hard to know what quality fresh meat should actually look like when it is right in front of you.

When you purchase meat from a butcher shop or supermarket, you want to know the meat you are buying is fresh. Here are some general guidelines:

- ✓ **Color:** Fresh meat should be "bright" and uniform in color without dark spots or dark edges. Note, however, that beef that's been dry-aged *should* be dark around the edges. Dry-aging enhances flavor and texture, but before you cook or eat it, trim away the dark, dried outer layer to expose the bright meat underneath. (Butchers will almost always tell you if meat is dry-aged, because a good aging program is a point of pride.)

  Many people wonder whether meat that's dark or discolored is rotten. The answer is no. When meat is dark, it simply means that oxidization has taken place. Darkened meat is more of an indicator of how long ago the meat was cut. For example, if you leave a fresh steak (cut today) in the fridge on a plate overnight, it will darken. Does that mean you shouldn't eat it? No, it just means you should have stored it better.

- ✓ **Smell:** Meat should smell mild and meaty (like blood). If the meat is bad, you'll know when you smell it. It will be pungent and smell . . . well, spoiled.

  Meat that is wet-aged in cryovac packaging (as is most of the meat we consume) has a strong odor when you open the package. That odor's okay; it's called *off-gassing*. After 15 minutes or so, the odor will fade away.

- ✓ **Feel:** To the touch, meat should be firm, not floppy, and definitely not tacky or slimy. Tacky, slimy meat is on its way out and would never be acceptable for sale in a retail environment. Don't consume it at home, either.

Identifying standard and specialty cuts

What is the difference between standard and specialty cuts? Standard retail cuts have been standardized by the meat industry. The industry accepts these cut names (like rib eye or Boston butt) and uses them to create a common language they can use for marketing and identification.

Specialty cuts are created more or less in house. This does not mean that specialty cuts are not common or won't catch on over time; it just means that, if I decided to name a meat cut at my butcher shop, although it may be recognized in my community, it is not standardized or recognized on a national level and wouldn't be used to sell meat in a wholesale business. It is a quirk of retail and restaurants that we sometimes give names to share our personality and creativity.

Substituting Cuts in Recipe Planning

Your favorite recipe is more flexible than you might think. Knowing how to interchange similar cuts is a real plus at the retail meat counter. Instead of worrying about a specific cut, try to think in broader terms, like flavors, textures, and cooking preparations. Say, for example, that your barbecue pork ribs are always a party favorite, but your local butcher shop has sold out of pork ribs. Don't panic! You can substitute a variety of other cuts for the ribs: pork butt, pork shanks, pork belly, lamb riblets, beef shortribs, or beef brisket, to name a few. All these cuts are fantastic braised or slow smoked in a sweet and tangy sauce.

Keeping your options open at the meat counter also helps you control how much you spend. If you can't afford to serve grilled rib eyes (a pretty pricey cut) with salsa verde and arugula, serve grilled skirt steak (a less expensive cut) with salsa verde and arugula instead.

In the following sections, I group cuts by cooking method to help you when you want or need to make a substitution.

Braising, slow cooking cuts

Braising cuts come from muscles that are tougher and typically have interconnective tissue. These cuts are best prepared over long periods of moist cooking. Slowly cooking these cuts in moisture allows the meat to break down and soften while the fats and interconnective tissues add rich depth and flavor to the dish. Here are a few examples of braising cuts that can be substituted for one another. Be sure to get cooking tips from your local butcher.

Brisket	Ox tail
Boston butt	Ribs
Breast	Shanks
Chuck	Shoulder
Neck	Stew meat
Osso bucco	

Grilling or quick-searing cuts

Grilling and quick-searing cooking methods don't provide much time for connective tissues to dissolve and the meat to tenderize. For that reason, cuts intended for grilling or searing should be tender and have good fat. They're best when served rare (cool red center) to medium (warm pink center).

The thickness of the steak should depend on how tender the muscle is. Very tender muscles can be cut into thick steaks and eaten in large, juicy chomps. Semi-tender, leaner muscles are better cut into thin steaks and quick-grilled rare or medium rare.

Here are a few cuts that are good grilled and seared and can be interchanged (remember to get cooking tips from your local butcher):

Flap meat (bavette)	Rib chops
Flank steaks	Rib eyes
Flat iron steaks	Sirloin steaks
Loin chops	Tenderloin
New York steaks	Top loin steaks

Roasting cuts

Roasting cuts have the widest range of tenderness and fattiness, so you need to know where they fall on the spectrum from tender to tough to know how to best prepare them. For example, pork butt and pork leg are on opposite ends of the spectrum, but both can be roasted to beautiful results. Here are some general guidelines:

- A roasting cut with a lot of fat can be slow roasted over a longer period of time, whereas a very lean cut can be brined and then roasted to help tenderness.

- To cook tougher cuts like eye of round to a delicious, finished product, roast it at a low temperature to medium rare and then thinly slice it (as you would roast beef).

- Properly salting the outside of the roast is very important. Unless you are brining or injecting your roast with a salt solution, the outside of the roast is your only seasoned surface. Use this as an opportunity to create a delicious, herby-salty crust.

Here are a few examples of roasting cuts and where, in general, they typically fall on the tender/fatty spectrum. If you've never cooked one of these before, ask your local butcher for guidance.

More Tender/Fatty	*Less Tender/Fatty*
Rib eye	Leg roast
Shoulder roast	Fresh ham
Cross-rib roast	Picnic roast
Loin roast (depending on the beast)	Top round roast

Chapter 4

Basic Knife Skills, Tools, and Techniques

. . . .

In This Chapter

▶ Selecting knives and other cutting implements

▶ Understanding the importance of the right grip and body posturing

▶ Getting familiar with special cutting techniques

▶ Paying attention to safety

. . . .

To get started butchering whole animals, you need a few things:

✔ **Butchery tools:** At the least, you need a few select knives, a cleaver, a mallet, and a bonesaw. Of course, other tools are available, which you may want to consider as your skill grows.

✔ **The right technique:** Butchery often involves standing in one position and performing repetitive tasks — work that can pretty quickly cause muscle strain and fatigue. In addition, you can own all the knives in the world and still struggle (or worse, hurt yourself) if you don't know how to use them properly.

✔ **Commitment to safety:** As a butcher, you work with very sharp implements and potentially heavy pieces of meat. To do so safely, you need to know the safety rules and follow them religiously.

In this chapter, I give you the details.

Knives, Mallets, and More: Gathering Your Butchery Tools

As with any project or task, starting with the right equipment is key. As someone new to butchery, you really need only a few things: a few knives, a mallet, a cleaver, and a handsaw. You can accumulate additional tools later to make your work easier and complement your skill set as you move forward in your training. But your initial investment shouldn't be financially overwhelming, so don't worry about buying expensive Japanese cutlery (unless you're a knife buff and want to). You can find good quality, affordable knives that stand the test of time and are commonly used and loved by butchers all over the world.

In the next sections, I list the knives you absolutely must have, some other equipment you shouldn't be without, and a few other items that, although not essential, can come in handy.

Take pride in the ownership of your tools. Keep them in the best possible condition: Don't skip things like oiling metal grinder parts after each cleaning or regularly sharpening your knives.

The essential cutting implements

As a butcher, your best friends and most-used tools are your knives, You use them to cut, divide, seam, bone out, hook, trim, and so on. Here are the knives you need to get started (you can see these in Figure 4-1):

- **Boning knife:** A boning knife has a sharp point and narrow blade (as opposed to a chef's knife, which has a thicker, wider blade and is used for chopping and slicing). Boning knives are designed for removing the bones of any kind of meat. Purchase a boning knife with a stiff blade; boning knives with flexible blades are used for fish, not meat butchery.

- **Cimeter:** A knife with a classic, curved blade, a cimeter is well suited for cutting a variety of meats from chicken to pork and perfect for cutting steaks and larger primals and subprimals. You can use the cimeter and the butcher knife interchangeably.

- **Butcher knife:** A butcher knife is designed for butchering and cutting beef. The blade works well for splitting, striping, and cutting meat, and this knife is commonly used in the meat processing trade.

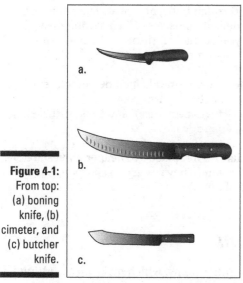

Figure 4-1:
From top:
(a) boning
knife, (b)
cimeter, and
(c) butcher
knife.

a.

b.

c.

Illustration by Wiley, Composition Services Graphics

You also need a *bonesaw,* which you use to cut through bone (professional butchers in busy shops often use a bandsaw, but as a home butcher, a hand-saw works just fine); a *meat cleaver,* which you use to break bone or chop meat; and *a rubber mallet,* which you use in combination with a large cleaver to break through thicker sections of bone. Figure 4-2 shows these three implements.

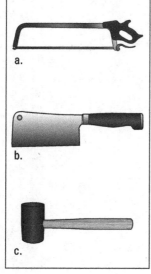

a.

b.

Figure 4-2:
From top: (a)
bonesaw,
(b) meat
cleaver, and
(c) mallet.

c.

Illustration by Wiley, Composition Services Graphics

Hand-held bonesaws are for sawing through bone, not meat. The blade of a bonesaw can do terrible damage to delicate muscles. When sawing through bone, make sure to stop as soon as you've made it through the bone and switch back to your boning knife to finish the task.

If you're looking to buy knives, some favorite brands include Forschner (http://www.cutleryandmore.com/forschner_knives.htm). Messermeister (http://www.messermeister.com), and Dexter Russell (http://www.dexterrussellknives.com).

You should sharpen a knife only so far, then replace it. Cheaper knives are typically made of softer steel, which is fine. They're easy to sharpen, but they don't hold an edge for a long period of time.

Other necessary items

Beyond your knives, you need a few other items, which make meat cutting a bit safer (see Figure 4-3):

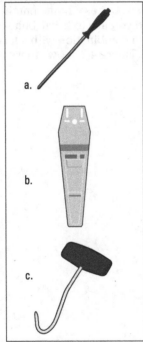

Figure 4-3:
From top:
(a) steel, (b) scabbard, and (c) boning hook.

Illustration by Wiley, Composition Services Graphics

✔ **Steels:** You use a steel to hone your knives by resetting "off" angles that develop on the blade's edge during cutting.

✔ **Scabbards:** This looks like a utility belt compartment except that it holds your knives and a steel on your hip for quick retrieval. Wearing a scabbard also eliminates the need to set your knives on the bench, which can be dangerous if you lose track of where you placed them and grab the blade by accident — bad news.

✔ **Boning hooks:** A boning hook functions as an extension of your working hand, keeping your fingers out of the way and providing a steadier grip than your hand can.

Useful but nonessential items

Here are a few items you'll find handy:

✔ **Bone duster:** You use a bone duster to remove bone dust from saws so that the meat doesn't oxidize on the surface. You can also use a towel or rag if you don't have a bone duster.

✔ **Tenderizer or mallet:** Different than the rubber mallet you use with a cleaver (refer to the earlier section "The essential cutting implements"), you use this kind of mallet to break up muscle fiber in meat, making it more tender. A tenderizer is mainly used for subprimal round and chuck cuts because they are lean and tough.

✔ **Meat grinder:** If you plan to make hamburger, sausage, or minced ingredients for sausage, you'll find a grinder and other sausage-making implements useful. Head to Chapter 14 for a list of sausage-making supplies.

✔ **Peach paper:** You use this paper to preserve and display meat, which needs oxygen to remain full and red. (*Note:* Some call it "steak paper"; others call it "pink paper.")

To stop your meat from oxidizing (and turning dark), place this paper between cuts of meat.

Making Confident and Fluid Cuts: Basic Grips and Posture

One of the things that makes butchers so impressive is how confident and powerful their cuts are. If you watch a good butcher in action, you'll notice that the motion of the knife is fluid and that the butcher works with, and not against, the weight and position of the carcass.

Pay closer attention, and you'll see that butcher's cut differently than other knife wielders, like chefs: Whereas chefs mainly use their wrists to cut, dice, and chop, butchers cut with their shoulders and keep their wrists in a locked position. Because shoulders have more strength than forearms, longer, more impactful strokes are possible.

Achieving the power and fluidity of the butcher's cutting technique requires that you use the correct grip on your knife and the proper body positioning. It also takes a lot of practice. But when you have enough practice, the knife and the cuts will feel like a part of your natural movement, and that's what you're aiming for.

Get a grip! Holding your knife properly

Using a wobbly or improper knife grip while cutting large pieces of meat can be dangerous. That's why understanding how to properly hold and wield a knife is very important. Sure, you may have been using a knife for years, but chances are you've picked up some bad habits. I've taught many butcher classes in my career, and whenever students' techniques have scared me, my fear wasn't that they were cutting the meat poorly but that they would cut off their fingers.

A common mistake inexperienced butchers make is putting too much force behind the knife while positioning their free hands too close to the unstable blade. Here are some general rules:

- ✔ **Don't force it!** You should not have to put excessive pressure behind your blade; a sharp knife should cut smoothly through the meat. If you have to push hard to get the knife through, stop. I can't stress this point enough: If it feels dangerous, it probably is.

- ✔ **Keep your fingers and hands away from your blade.** This is one reason why you want to use boning hooks.

  When you're cutting meat, your hands are often slippery and/or cold, which makes it that much more important to avoid putting your fingers near the blade or putting undue pressure on the knife.

- ✔ **Take it slow.** Never be in a hurry. Go as slowly as you need to, to butcher safely, and take the time to develop good, safe skills and techniques.

Peter Hertzmann, author of *Knife Skills Illustrated: A User's Manual,* explains how to properly hold a knife: "The grip you use in butchering is both a function of what you are cutting and the direction you are cutting it." Following is his advice:

✔ **The pinch grip:** Most of the time, when you are cutting meat against your cutting board, you hold the blade vertical and slice back and forth. In this position, a pinch grip provides you with the most stability. You also use the pinch grip when cutting horizontally with the cutting edge of your knife pointed toward your holding hand.

To grip any knife with a pinch grip, grasp the blade with your index finger and thumb right in front of the handle, or choil if the knife has a bolster, and firmly hold the handle with your remaining three fingers. To avoid morphing this grip into a granny grip, don't extend your index finger onto the spine of the knife.

The granny grip is appropriate when cutting fish with certain Japanese knives, but it has no place in the butchery of meat or poultry. It is inherently unstable and can cause tendonitis over the time.

✔ **The thumb grip:** You use this grip when you use your knife to cut in the direction away from your holding hand. This thumb grip is good for removing silver skin from loins, but be aware that long-term use of this grip can cause bone spurs in your thumb, especially if you have arthritic tendencies.

To grip the knife with a thumb grip, roll your hand laterally a quarter turn around the handle. Grip the handle firmly with your four fingers and place your thumb directly on top of the spine for leverage. Don't extend your thumb over the blade.

✔ **The dagger (or butcher's) grip:** When cutting in situations where the cutting edge is directed towards the floor (as when you're cutting downward on a hanging carcass) or when the tip of your boning knife is pointed towards your cutting board (as when you're hand-chining ribs), hold your knife in a dagger grip. (I prefer to think of this as the *Psycho* grip, but that won't mean anything to you if you haven't seen the 1960 Hitchcock movie.) This is a full-handed grasp of the handle with your thumb towards the butt of the knife and the cutting edge directed at your forearm.

Peter continues, "Here's a way to keep the grips straight: Use a pinch grip when you're slicing the meat. Use the thumb grip when you're pushing your knife through the meat. Use the dagger grip when you're pulling your blade down through the meat." Figure 4-4 shows the different kinds of grips.

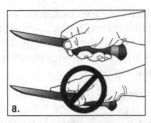

Figure 4-4:
From left: (a) the pinch grip, (b) the thumb grip, and (c) the dagger grip.

Illustration by Wiley, Composition Services Graphics

Maintaining good posture

Butchery involves a great deal of focused, repetitive work. I endured a debilitating back injury several years ago (as a result of poor lifting) when my business partner and I first opened the butcher shop. Basically I lifted something the wrong way and paid the price. Make sure you take your posture and movement seriously.

When you butcher, you'll find yourself standing for long periods of time at a cutting table. To avoid back pain, make sure the table is a the right height; wear good shoes; lift with your legs, not your back, and don't twist while lifting; and consider standing on a mat rather than the hard floor.

Marissa Guggiana, author of *Primal Cuts, Off The Menu* and co-founder of the Butcher's Guild offers this advice for avoiding back pain: "After setting up your work area so that you're cutting at a height where you aren't stooping, the next best preventative step for avoiding back pain is to pay attention to your posture. Employ a powerful stance, distributing your weight equally over both feet with a very slight bend in your legs. This stance lets you move from your legs and core without torqueing your back. Finally, take stretch breaks.

Open your chest and rotate your arms to counteract the forward contraction of butchering. Pull back on your hands, one finger at a time, with your palm facing away from you. Touch your toes. Your body will give you some cues about what will be sore later, so be proactive and give those muscles some attention."

Special Techniques Every Butcher Should Know

Holding the knife properly and having the right tools are definitely integral components to your success at butchery, but you need to know a few special techniques as well. If you're a home butcher, these tricks let you impress your friends and family with the professional quality of your cuts, and if you're a butcher in a shop, they're skills your customers (or boss) will expect you to know.

Professional butchers have a saying: "Meat that is ready for the case is meat that is ready for plate." So one goal of your butchery endeavors is to create cuts that work with standard cooking methods. For example, if you're cutting short ribs, the ribs should be uniform in size, trimmed of silver skin, attractive, and so on. If you're cubing meat for stew, you want to make sure that the cubed stew meat is uniform in size and trimmed of inedible connective tissue (gristle) or extraneous fat and silver skin. A poorly cut steak isn't good marketing for your business, and it doesn't look too great on a plate, either.

Retail cuts *should* look attractive, but most importantly they should be edible. Cutting against the grain, denuding, cubing meat for stew or grind, butterflying, and frenching chops are skills you'll want to have because you'll use them often when you cut down a carcass.

In the following sections, Butcher's Guild member and butcher Pepe offers tips on fabrication techniques from her experiences butchering at Marlow and Daughters in New York.

Denuding

Denuding means cutting away all fat and sinew/silver skin from a piece of meat. In the shops where I have worked, we denude all of our steaks before they go into our retail case.

Cutting away the excess fat

To denude a piece of meat, you start by cutting away all of the excess fat from the exterior of whatever piece of meat you are preparing. Don't cut too deep, or you will cut into the meat itself, which is a waste. Cut the layers of fat away. When you start seeing red meat underneath the fat, stop cutting. This is as far as you want to take it. If you want to leave some fat on the piece of meat to enhance the cooking, this is also fine, but trim it to a reasonable amount.

If the fat is good and hard and hasn't been exposed to anything that would make it unusable, use the fat for grind (ground beef for burgers and such). If the exterior that you have just trimmed has been exposed to anything that would contaminate it, then toss it in the compost. Some of the exterior of beef becomes hard and is no good for anything after you've trimmed it off. Once you have trimmed the fat, the next step is to remove any silver skin or sinew.

Removing the silver skin

Silver skin is a butchery term referring to a type of connective tissue found in most animals. The name should give you a clue: To find the silver skin, look for a thin membrane with a silvery sheen.

To remove the silver skin, insert the tip of your boning knife underneath the silver skin and put a bit of pressure against the silver skin to peel it away from the meat. Do this to all of the silver skin present. Discard this silver skin; it is of no use to you.

If large pieces of sinew are on or in the meat, you have to cut that away, as well. After you have done this, you should be left with a lovely piece of meat that is ready to be cut and cooked!

Cutting steaks

To cut a nice steak, drag your knife cleanly and quickly through the length of the muscle and cut each steak into neat squares or rectangles (or whatever shape you can get out of it). To do this task, use your free hand to hold the meat on the opposite side of where you will be cutting while you make the cut. This technique allows you to have the most control over the larger piece of meat that you are turning into more "steakable" pieces.

After you have cut all your steaks from the larger muscle and they are all about the same size, you can trim each steak to make it as uniform as possible. Cut away any bits that make the steak uneven. Doing so makes the steak look more appealing and allows it to cook more evenly — which is especially important for smaller steaks that require only a few minutes of pan-searing on each side. After you've cut away any pieces necessary to make the steak cook evenly, you can reserve those pieces to use later in a stew, stir-fry, or any other preparation that you want.

Frenching

Frenching a piece of meat means exposing the tip of the bone for a more elegant presentation and even cooking (think of a crown roast or a rack of lamb). Basically, in frenching, you cut and scrape the meat away from the tip of the bone to expose the bone. As a home cook, you'll probably french a chop at a time (Chapter 12 has instructions for frenching a bone-in rib eye). You can also french a rack at a time. Follow these steps:

1. **Place the rack with the rib bones down flat on your cutting board, and mark the spot on both sides of the rack with a small slit to note where you'll make your first cut.**

   Looking at the rack from the side, make note of where the eye of the loin is and where the small, triangular piece of fat above the eye comes to a point. You'll be cutting about 1–1½ inches above this triangular piece of fat.

2. **Cut lengthwise across the entire rack from slit to slit, being sure to cut all the way to the bone.**

   Flip the rack over (you will now be looking at the rib side of the rack) and, using the tip of your knife, make a cut on the membrane from the top of the rack where the rib bone ends down to the your previous cut mark on the other side of the rack.

   The knife will easily cut through the membrane, and you'll be scraping the rib bone with the tip of your knife, which is what you want. You are essentially scraping the membrane away from the bone in order to allow yourself to pull the piece of meat/fat off of the tip of the rib rack to expose the clean bone. After you scrape the membrane from each rib bone, you'll be able to pull the piece of meat off the rack along the line that was your first cut.

3. **Cut the meat in between each bone, making sure to fully expose each rib bone.**

   Doing so lets you pull the meat away.

4. **Flip the rack over again and with your hands, start pulling away the 1–1½-inch piece of meat away from the entire rack, leaving the bones clean.**

   If some bits of meat have remained attached to the bones, you can scrape the meat off of each rib bone separately, or rub it off with a clean kitchen towel.

Butterflying

Butterflying means filleting a large piece of meat open, along seam lines, in order to have an even and thinner cut of meat. Butterflying can be a useful technique for roasting a larger cut of meat, such as a lamb leg, or when you want a larger cut of meat to cook more evenly and/or more quickly.

After butterflying a large piece of meat, season the inside with salt, pepper and/or herbs, and then roll the roast up and tie it so that it cooks evenly.

To butterfly a piece meat, follow these steps:

1. **Keeping the meat flat on your cutting surface, place your knife on the outside portion of the meat and cut horizontally through the meat while pulling the meat back with your free hand. Do not cut all the way through the meat to the other side.**

   The idea is to flatten the meat out even further and to end up with one, solid piece of meat that is the same thickness all the way through.

2. **Repeat Step 1 on the other side (the right side of the meat) to flatten out the right side, but don't cut all the way through.**

When you're done, the large piece of meat should be relatively flat and much wider than when you started and should be the same thickness throughout, allowing you to season it and roll it up for roasting.

Merchandising and retail meat cutting

Cutting meat at home or as a hobby is a whole different beast than cutting meat in a retail setting. Meat is your money and "the steaks are high" (pun intended).

Josh Martin, from Whole Foods, has worked in the meat industry for 11 years. His experience covers many facets of the industry: major supermarket chains, warehouses, local chains, and straight-up butcher shops. He shares his thoughts and tips on merchandising and working at a retail meat counter.

On handling the meat case:

"When handling meat in a retail environment, you have to use your senses, as well as your overall knowledge of the industry and customer expectations. Customers mostly judge a purchase initially on sight, so as a butcher I think of myself as my own customer. I will not put out a piece of meat if it is oxidized, smelly (ammonia or citrus smelling), rotten, jaggedly trimmed, uneven, broken wrong, overly fatty, has water burns, or is not appealing to the eye.

"On occasion though, you will find scarring in the muscle tissue, in particular with the beef round cuts, due to mishandling of the animal, bad fencing or roping, and tussles out on the pasture. In that case, you must trim or remove the scarring to maximize the quality and appearance of the final cut.

"I also look for meat from happy animals that were allowed clean conditions and pasture because it reveals beautiful cuts. An animal that was stressed or abused can produce an unnaturally colored, textured, and flavored meat. Meat that is pale, soft, and exudative (PSE) usually has little or no flavor. PSE is most common with animals that have literally no wiggle room and are force-fed. Meat that is dark, firm, dry (DFD) usually comes from animals that are abused or from uncastrated male animals, which produce a lot of adrenaline and burn up glycogen stores. This causes the meat to be tough, sticky, and strong in flavor."

On handing the tools in a retail meat shop:

"You must be aware of safety. Many common mistakes are made when cutting meat in a high volume retail setting. If you want to keep all of your extremities, be aware of these mistakes so that you can avoid them and know that safety is paramount when cutting meat at a fast pace:

- Leaving knives under towels, trays, blocks, and other things. If you aren't careful, a severe cut is bound to happen — I know from experience.

- Using dull knives. It's much easier for the doc to sew up a clean cut rather than a jagged one made by a dull blade. Jagged cuts can also produce nerve damage.

- Wearing gloves while using bandsaw. A bandsaw will literally catch the glove and pull your hand into the blade.

- Not wearing steel-toed shoes. Much of the work and equipment used in the meat industry is extremely heavy and can crush toes. Think about frozen turkeys. Who wants to catch a 12 pound bird with their foot?

- Not wearing a cutting glove. Although some very experienced butchers opt not to wear cutting gloves, these gloves are designed to protect you from cuts."

On showcasing your cuts:

"Making money is the most important thing in retail, and to make money butchers have to find every means available to sell a very perishable item at maximum retail margin. Strategies include using the whole animal (organs, bones, fat, blood, and anything else they can sell), creating signature products (sausages, marinades, meat rubs, and so on), and creating displays that attract customers. Consider using garnishes, like herbs and lemon wedges, or tying your roasts with sprigs of fresh rosemary. Build beautiful trays: Use colored paper, play with the placement and organization of the meat in the case, and more. Take notice of customer feedback and create a signature look for your case. After all, your case is your identity and the face of your business.

"It is also important to know and understand food trends in order to cater to your customers. Ten years ago, for example, no one would ever pay almost $15 a pound for a marrowbone and crostini appetizer, but now that item is somewhat of a staple in meat-centric restaurants. If restaurant goers enjoy it enough, they want to re-create it at home and will go to the butcher to purchase it. That's why it's important for me to be able tell them how to prepare it. The knowledge I have of the food industry can help me build a steady customer base. Butchers are a healthy part of local food systems, providing the necessary steps for someone to become a great home chef, and I'm proud of it.

"The most important part of building a successful retail case is knowing your customers, knowing what they are willing to spend, knowing their culture and cuisine, knowing modern food trends and fusions, and knowing how to merchandise every part of every muscle of every animal. These steps have helped me in providing the best retail case possible for my customers. And have fun with it. Finding new ways to sell and market cuts is one of the most rewarding challenges of the job."

Cubing meats for braising

When you are cubing meats for braising, you want to make sure that your pieces are uniform in size so that they cook at the same rate and are the correct size for the preparation. If you are making a beef stew, for instance, you want each piece of meat to be bite-sized so that no one has to hack away at it with their spoon in order to take a bite. If you are making braised chicken, however, you would assume that all diners would get a piece of chicken on their plate and that it would be okay to braise the chicken in larger pieces.

To cut pieces of meat for braising, take your chef's knife or your boning knife and cut strips of meat from a larger piece, if you are starting with that, and then cut each strip into the size cubes you've deemed appropriate. Try to make each piece the same size so that all the pieces cook at the same rate.

Being Safe While Using Sharp, Pointy Metal Tools

Sharp, pointy metal tools like knives need special consideration because they have the potential to cause serious injury in the blink of an eye. Take time to establish safe, healthy habits on the bench so that you can avoid injury to yourself and those around you.

Josh Donald, from Bernal Cutlery, offers a list of good habits to remember:

- **Keep your knives sharp.** Dull knives might seem more benign and less eager to bite the hand that wields them, but in fact they require much more force and have a bad habit of slipping off their intended track. Sharp knives will obey your intentions far better.

- **Don't have a secret knife.** Let others know about your knife when walking in a workspace: Hold your knife down and say "knife" in a clear voice when walking past. When moving around those using knives, announce your presence by saying "behind you" or "beside you." Don't turn around with your knife facing out. Don't put your knives in soapy sinks or piles of dirty dishes, where they will be hidden and accidentally grabbed. Wash your knives and dry them as soon as you're done using them.

- **Use a knife only for its intended purpose.** Don't use a knife meant for precision work to hack or a knife designed for slicing to poke into a tough item. And please don't use a knife for cutting plastic clamshell packaging — the single most likely way that you are going to introduce yourself to your local ER team. It's not a can opener or screwdriver, you gorilla.

✔ **Don't catch a falling knife.** You just spent $300 on a hand-made Japanese knife that is your new pride and joy, and it nudged its way off the counter. You are going to watch it fall and then inspect it for damage. You are *not* going to catch it because you can get another knife more easily than you can get another hand. Plus you'll spend more than $300 on your ER excursion anyhow.

✔ **Resist the temptation to play the knife ninja or practice your spring break bartender moves.** A knife is a tool, not a toy. Do not employ the flaming-sword-of-glory technique to hone your knife with a sharpening steel. That's a great way to get a knife wedged in your knuckle.

✔ **Work only as fast as you can safely work.** Those who have fast knife skills worked their way up to that speed. Haste makes great stories about how many stitches or severed nerves were earned. By working with focus, you can avoid inadvertently putting your hand or other body part in the path of your knife.

✔ **Wipe your knife with a double- or quadruple-folded cloth.** Do not use a single layer of cloth because a knife can easily cut through it.

✔ **Don't experiment with power tools for your sharpening needs.** Avoid using fast bench grinders. Not only do they ruin your knives, but they can also quickly grab your knife and throw it back at you. Consider it cutlery karma: If you take shortcuts sharpening your knife, your knife may shortly cut into you.

✔ **Keep your knives covered when not in use.** Don't leave them loose in a drawer, bag, box, and so on. Covering them not only protects your knives from dulling, but it also protects you.

✔ **Consider using cut gloves.** They're inexpensive and easy to find. You can order them online or purchase them at a restaurant supply store. They are machine washable so you can use them over and over. There is nothing tricky about how to use them, just put them on and start cutting!

Not only do knife cuts hurt you, but they can also inhibit your work. Normally, a cut requiring stitches takes from two to six weeks to heal, during which time you may not be able to work. You also risk the chance of long-term damage to your hand. Your hand has many tendons, and a deep cut could easily be enough to catch a tendon, which will effect the use of your hand, possibly forever.

Part II
Poultry, Rabbit, and Lamb Butchery

The 5th Wave   By Rich Tennant

"You make such a procedure out of boning a chicken."

In this part . . .

If you want to get started butchering, starting at the beginning is best, right? And the first thing you need to do to be a good butcher is to develop strong, comfortable knife skills. By spending a good amount of time butchering the smaller beasts like poultry and rabbit, you can develop the skills you need to move on to the bigger beasts like lamb and goat (covered in this part) and pork and beef, covered in Parts III and IV.

Chapter 5

Duck, Duck, Goose, Chickens: Starting with Poultry

*T*his chapter is about birds: big birds (geese, ducks, turkeys), little birds (quail and squab), and birds (think chicken) that have different names, like hen, roaster, poussin, and others, depending on their size. In general, poultry butchery applies to any avian species with similar body structure. If you want to become a more skilled butcher at home (or are even considering becoming a professional butcher), poultry makes a great starting point.

Although turkey, duck, chicken, and game birds can be roasted whole, they can also be broken down into parts, sectioned, or deboned. Because the chicken is the most universally eaten of all meats (in addition to being versatile, which lends it to a glorious variety of preparations), the butchery instructions in this chapter are for, but not limited to, chicken. Here, you can find out how to cut up an eight-piece fryer chicken; a boneless, skinless chicken; and a brick-style bird. I also offer a few extra tricks you can use to impress your friends and relatives. After working through each of these techniques a few times, you should feel comfortable enough using your knife, your fingers, and the bird itself to butcher in a natural, flowing way.

Butchery is also a skill that requires practice. If you're butchering at home, remember that, as with most complicated things, practice makes perfect. Don't be disheartened if you don't get it right the first time. Try, try again! If you're an apprentice butcher, make time to cut your way through many pounds of chicken before you ever touch a hanging carcass.

Going for farm-raised heritage breeds

Tenderness and flavor are greatly affected by the life of the bird before it reaches the butcher's case. The very best tasting birds are heritage breeds; heritage or heirloom breeds are traditional livestock breeds that were raised by farmers in the past. Industrial farming has reduced the number of diverse breed varieties available, but small farmers are bringing back heritage breeds, allowing us to taste the benefits of breed diversity once again. (Interesting info about heritage birds: They have different proportion of breast meat to leg/thigh than do commercial breeds, and they also retain the ability to fly because they don't have out-of-proportion bodies.)

Fortunately nowadays, you have more options for local, farm-raised birds that have much better flavor than the tasteless commodity chickens we've all grown accustomed to. So buy locally raised birds. It's a good way to support a healthy local food system and your community. And if you can, try a heritage bird.

A Word about Cutting Up Birds

The term *poultry* refers to any type of domesticated, edible bird: duck, turkey, chicken, quail, and so on. In the United States, the most common type of poultry is the chicken (no surprise there). In fact, worldwide, nearly 200 varieties of domestic chicken exist.

A *game bird* is any bird that is hunted for food or sport. As is common with most domesticated "wild" animals sold for food, game birds are farm-raised in conditions that mimic their wild habitat.

Tenderness and flavor are greatly affected by a bird's treatment before it gets to the butcher or market. When shopping for the best tasting bird, keep these points in mind: One, the fresher the better. Two, freezing always affects a bird's texture and taste. Three, chicken falls into categories, based on its age and size. Table 5-1 has the details; it also lists the other kinds of poultry you may find yourself working with.

Table 5-1	Types of Poultry		
Name	*Size*	*Age*	*Notes*
Quail	6 ounces	6 weeks	Sold whole
Squab	¾–1 pound	1 month	Sold whole
Poussin (chicken)	Small, 1 to 1½ pounds	3–4 weeks	Sold whole
Cornish game hen	Small, 1–2 pounds	4–5 weeks	Sold whole
Broiler/fryer (chicken)	Midsize, 2½–4½ pounds	6–10 weeks	Sold whole or in parts
Roaster (chicken)	Large, 4–7 pounds	9–12 weeks	Sold whole or in parts
Capon (chicken)	Large castrated male 6–9 pounds	9–12 weeks	Sold whole
Hen/stewing fowl (chicken)	Older, large female chicken, 4½–7 pounds	Typically more than 10 months	Sold whole
Duck	4 pounds	2–6 months	Sold whole or in parts
Goose	6–14 pounds	12–14 weeks	Sold whole
Turkey	5–25 pounds	16 weeks (fryer) 15 weeks (roaster)	Sold whole

Getting familiar with poultry musculature

A joint connects two or more bones, ligaments connect the bones to each other and anchor the muscle to the bone, and the muscle holds everything together. As you work with your chicken or other poultry, try to get a feel for the joints and follow the muscle divisions. These are the kinds of joints you'll encounter:

- ✔ Ball-and-socket joints, like shoulder and hip joints
- ✔ Hinge joints, like elbow and knee joints
- ✔ Pivot joints, like the neck joint which enable twisting motion.

When you butcher, use these skeletal functions to find the *cutting points,* places where you can easily and efficiently remove pieces of the carcass. Move the joints around if you feel unsure about where to make your cut. By paying attention to how the bones move, you can see where the ball joint is connected to the socket. As you manipulate the joints, use common sense. Bend a joint in the opposite direction of its natural rotation, and it'll pop right out. For detailed information on animal musculature and how to use it to your advantage while butchering, refer to Chapter 3.

Basic chicken-butchering tools and techniques

For butchering any kind of animal, a trusty boning knife and a sharpening steel are your most trusted tools (refer to Chapter 4). You may be more familiar (if you're a home cook) with using kitchen shears or a large chef's knife to cut up chicken, but have no fear! A boning knife is your new best friend, and the best tool for the task.

Follow these important tips for safe and successful chick butchery:

- ✔ **Always use a sharp knife.** Just because the animal's small doesn't mean you don't need a sharp knife. Dull knives have been the culprit of many a flesh wound. Chapter 4 discusses knife safety.

    You may choose to wear cut-resistant gloves to help you get a better grip while you are learning, but accidents still happen, so keep a sharp eye on that sharp blade.

- ✔ **Make sure your chicken is thoroughly thawed before you cut it.** If you're processing a chicken that has been frozen, plan ahead: Let it defrost completely beforehand. The best way to thaw meat is in the refrigerator over night.

    Didn't plan far enough ahead? Here's a safe way to quickly thaw chicken: If you're in a pinch, you can thaw frozen chicken under cool running water until it is defrosted. (I don't recommend defrosting meat in the microwave.)

- ✔ **Discover how the bird lays most steadily on the table.** Move it around to find a stable cutting position.

- ✔ **Use the tip of your boning knife to slice close to the bone, making small cuts and using your hands to help pull the meat away as you go.** Take a look at Chapter 4 for more information on what you can do with a boning knife.

Pieces of Eight: Cutting Up a Fryer

Fryers, or in industry terms, WOGs (short for "With(O)ut Giblets") sell standard as whole animals. Where better to begin your foray into butchery than with cutting up a whole chicken? When you butcher a fryer (referred to as "whole chicken cut up"), you cut an entire bird into eight individual pieces: two breasts, two wings, two thighs, and two legs. (If you purchased a farm-raised bird, you may also have a head and two feet.)

You may think, "*Please,* anyone can cut up a chicken. That's not real butchering." On the contrary. You'd be surprised at how many people don't know how to cut up (or cook) a simple fryer chicken any more. The eight-piece fryer is a butcher classic! And the cuts you get from it are great braised, baked, grilled, and, of course, fried.

In the following sections, I take you step by step through the process. As you proceed, try to get a feel for the movement of the carcass and use your hands and knife together to section the chicken into pieces efficiently and without waste.

Removing the head and feet

If you purchased a local, farm-raised bird, it will come to you with the head and feet still attached. You want to remove these extremities first. If you bought a chicken from the meat counter or meat case in a grocery story, the head and feet have been removed for you; you can jump straight to the next section.

To remove the head and feet, follow these steps, shown in Figure 5-1:

1. **Position your cimeter (or larger, curved knife) at the base of the neck (Figure 5-1a).**

2. **Use the palm of your hand to deliver a quick whack to the top of the knife, using blunt force to split the neck vertebrae.**

3. **Make one more solid cut to sever the neck all the way through (Figure 5-1b).**

   Set the neck and head aside for stock (you also want to save any fat and trim) and move on to the feet.

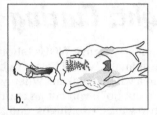

Figure 5-1:
Removing
the neck.

To remove the feet, take a look at the joint below the drumstick and find the lowest joint; using a boning knife, cut through the joint to remove the foot (see Figure 5-2). Do the same thing with the second foot. Add the feet to your growing stock pile.

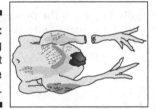

Figure 5-2:
Removing
the foot
below the
drumstick.

Stock is the perfect start to soups and stews, rice, stir fry, sautéed veggies, and almost any other dish that needs a little moisture. And of course, a cup of hot, rich, flavorful stock on a cold winter day is a nourishing, warming treat. So don't forget to take stock seriously and reserve the time to make it

Removing the wings

After removing the head and feet, your next task is to cut off the wings. When removing the wings, use the weight of the chicken body to assist you.

Skilled butchers don't merely know how to use knives to make the necessary cuts. They also use the weight of the meat and gravity as leverage to get the job done.

To remove the wings, follow these steps, shown in Figure 5-3:

1. **Grab the wing, much like a handle, and pull it up towards your body, away from the rest of the chicken, applying pressure to open up the joint (Figure 5-3a).**

   Extend the arm as if the bird were raising its wing up in a full stretch. The legs should be facing away from you.

2. **Lift the chicken by the wing off the table (the body should hang down a little) and make a shallow slice through the skin and into the meat.**

   This cut allows you to see inside so that you can make the next cut more easily.

3. **Cut further down through the ball joint inside the armpit (Figure 5-3b).**

4. **Being careful not to cut into the breast meat, use your knife to cut under the wing joint and back up and around the top of the arm (Figure 5-3c).**

5. **Now take the wing off completely by cutting through the flesh and removing the wing from the carcass (Figure 5-3d).**

6. **Turn the bird over and repeat Steps 1 through 5 on the other side.**

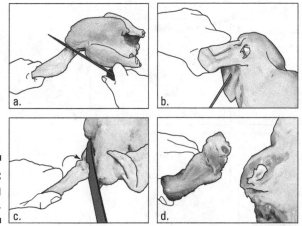

Figure 5-3:
Removing
the wings.

Illustration by Wiley, Composition Services Graphics

Removing the legs

After removing the wings, you remove the legs. With the legs toward you, turn the chicken on one side and follow these steps, shown in Figure 5-4:

1. **Examine the leg to identify your cut mark (Figure 5-4a), and with your knife, slice through the open space into the skin to mark where you'll break.**

   You can see quite clearly the space in between the skin and the leg. That's your cut mark.

2. **With one hand on the breast and the other on the leg, open up the leg, turning it outward (Figure 5-4b).**

   This move exposes the thigh bone so that you can see where it's connected to the pelvis.

3. **At the bottom of the leg, make a cut straight down until the knife runs into the thigh bone; then use your hand to turn the leg out and expose the joint, popping the joint out of its socket (Figure 5-4c).**

   Notice where the thigh bone was attached to the cartilage at the pelvis before you disjointed it.

4. **With one hand holding the knife with the blade facing away from you and anchored under the cartilage, and your other hand holding the top of the leg, pull the leg away from the body (Figures 5-4d and e). Cut through the skin to remove the leg completely.**

   If everything's gone according to plan, the meat neatly rips away from the bone, leaving the oyster (the small, round piece of dark meat near the thigh) intact and the bone clean. In other words, all the meat ended up where it should be, and each muscle remains whole.

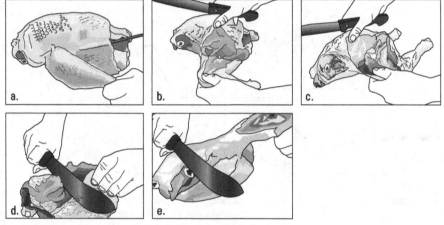

Figure 5-4:
Removing
the leg.

Illustration by Wiley, Composition Services Graphics

5. **Repeat Steps 2–4 on the other side.**

Voila! Both legs are removed.

Cutting out the spine

With the legs gone, you have an open view of the back and the breast. You are ready to make easy work of removing the spine. Follow these steps, shown in Figure 5-5:

1. **Take the leg end of the spine with your free hand and use it as a handle to hold the chicken upright.**

   This nob is called the *pope's nose.*

2. **With the knife in your other hand, start at the tail end of the bird and cut at a 45 degree angle down and into the spine (Figure 5-5a).**

3. **Turn the chicken over and look at the back.**

   Identify two small wishbones, one on either side that you will use as a cutting guide (Figure 5-5 inset). You want to make sure you keep your cut on the inside of that bone (in between the spine and the wishbone.)

   If you cut above the wishbone into the breast, your knife will get hung up on the bone, and you'll have to break through it, which wastes breast meat and produces a sloppy-looking cut.

4. **Push your knife through a few ribs, making sure your cut ends up between the spine and the wishbone (Figure 5-5b).**

5. **Finish by cutting straight down (Figure 5-5c).**

   Your knife will easily come out with one side of the spine completely separated from the breast.

6. **Turn the bird around and do the same thing on the other side to completely remove the spine from the breast (Figure 5-5d).**

   Save the spine for your stockpile.

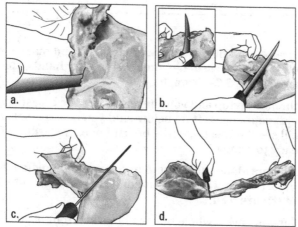

Figure 5-5: Removing the spine.

Illustration by Wiley, Composition Services Graphics

Butchers make decisive cuts, using their shoulders and not their wrists to do the work. This cutting style works best, especially when you're trying to get though thin or awkward bones such as a chicken back. To adopt this style, hold your wrist in a locked position and let your shoulder strength move the knife down through the meat. But be vigilant of the rest of your body, especially your fingers, when doing so.

Splitting the breast

With the spine gone, the time has come to split the breast. Take a look at the bird: You'll see the two wishbones sticking out that you cut around when you removed the spine (refer to Figure 5-5). Make note of those and then follow these steps to split the breast, shown in Figure 5-6:

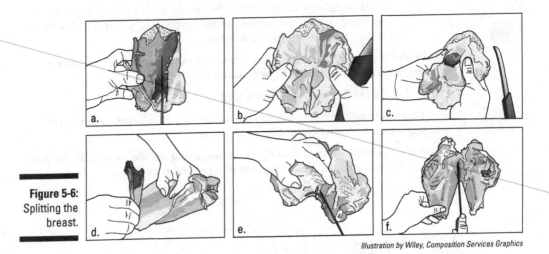

Figure 5-6:
Splitting the
breast.

Illustration by Wiley, Composition Services Graphics

1. **Turn the breast with the wishbones facing you and make a cut directly in the center, into the cartilage; then, using your hands, split the breast open (Figure 5-6a through c).**

   Put your thumbs on top and your fingers under the breast and then push up with your fingers from below until you hear a crack. After that crack, you'll see the keel bone (the breast bone, or sternum) is pushed out and ready to remove.

2. **Run your fingers along the silver skin to expose the whole bone; then hold the breast down with one hand and rip the bone out with the other hand (Figure 5-6d).**

   At this point you have a whole bone-in breast in front of you.

   Silver skin is an opalescent connective tissue attached to various pieces of meat. In chicken, you'll notice a lot of it when you split the breast. Because it doesn't offer any benefits when you cook or eat the meat, go ahead and remove it. Using the tip of your knife, make small cuts where the silver skin connects to the breast tissue and pull it away.

3. **Remove the wishbone by slicing around it with the tip of your knife (Figure 5-6e). Use your fingers to pull it off of the breast.**

Note: Removing the wishbone makes the last cut (Step 4) easier. Of course, if you prefer, you can leave the wishbone in and simply cut through it in Step 4.

4. **Use your knife to cut the whole bone-in breast right down the middle, separating the breasts in two (Figure 5-6f).**

Easy enough, right?

Dividing the legs into two pieces

The last part in the eight piece puzzle is dividing the whole legs into their piece parts. You want to simply cut them into two pieces, separating the thigh from the drumstick. Once again a guiding line of fat comes in handy. Follow these steps, shown in Figure 5-7:

1. **Find the line of yellow fat (Figure 5-7a).**

   This line of fat is where the joint is; using it as a guide is an easy way to feel confident that you know where to make your cut.

2. **Make the cut right on top of this line of fat (Figure 5-7b).**

   Cut through the joint with the line of fat as your guide ensures that your knife will glide easily through the joint, without resistance, for a satisfying, efficient cut.

Figure 5-7:
Dividing the
leg into two
pieces.

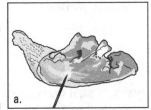 

a.　　　　　　　　　　b.

Illustration by Wiley, Composition Services Graphics

Finishing up

With the cuts outlined in the previous sections, you now have two bone-in breasts, two wings, two thighs, two drumsticks, and a pile of bones you can use for stock. You may choose to do some additional cut work; it's entirely up to you. Common finishing cuts include the following:

- ✔ **Partially removing the exposed rib bones on the side of the breast:** These final cuts are purely cosmetic, but that is what butchery is all about: making meat look good. So take a moment to remove any exposed rib bones with a cimeter or chef's knife.

✔ **Cutting off the wing tips:** You may also choose to remove the wing tips. A small, easy cut at the point where the wing bends from the arm to the tip gets the job done.

✔ **Cutting the breast in half:** Doing so creates more like-sized pieces for consistency during cooking times.

Cutting the Chicken into Five Equal Portions

In this section, I explain a second way to prepare a whole chicken: cutting the fryer into five equal portions. Dividing a chicken this way may seem a bit out of the ordinary — after all, you're not likely to find a chicken divided this way in a butcher's case. But if you're a wing fan, these cuts create a tasty bonus because of the unique distribution of the breast meat. When you cut a chicken into five equal portions, you end up with wing sections that include breast meat.

This technique produces five larger, similar-sized portions, and it's geared more toward the home cook because the equal pieces are perfect for stewing, grilling, or sautéing. Plus, it lets you utilize the whole bird, using as much of the meat as possible and leaving more of the bones on the bird and out of the stock pot.

In the following sections, I explain how to make the necessary cuts to get your five equal portions. When you're finished, you'll have two full legs, two wings with breast meat, and one full double-breast portion.

Freeing the oysters

Bet you didn't know that a chicken has oysters. Well, it does. Oysters are two small, round pieces of dark meat on the back of poultry near the thigh. The French word for oyster roughly translates to "the fool leaves it there." And, indeed, leaving it there when you're cutting your chicken into five equal portions would be such a waste! To free the oysters, you need to make a cut down the back.

If your bird still has its head and feet, remove those as I describe in the earlier section "Removing the head and feet"; then follow these steps to cut the back and remove the oysters, as shown in Figure 5-8:

1. **With the chicken on the cutting board, breast side down, make a 3-inch slit across the back below the shoulder blades (Figure 5-8a).**

   If you're not sure where to make this cut, use your fingers to feel the back and locate the ends of the shoulder blades.

2. Starting at the center of the slit you made in Step 1, cut down the center of the backbone, finishing right below where the hip joints are.

Be careful not to slice into the backbone. You're not trying to cut the piece free in this step; you're merely scoring through the skin to create a cut line. This cut preps you for removing the round, meaty oysters, which nestle snugly in small hollows on each side of the backbone.

3. With the tip of your knife, free each oyster from the bone by making small slices against the side of the spine and into the small of the back where the oyster lies (Figure 5-8b).

Make sure that, as you slice, you keep your knife against the bone. Your goal is to free each oyster from the bone but leave it attached to the skin.

Figure 5-8:
Freeing the oysters.

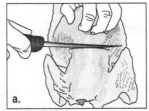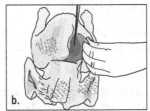

a.   b.

Illustration by Wiley, Composition Services Graphics

Removing the legs and spine

After you free the oysters from the bone, you can remove the legs and cut out the spine. The trick in removing the legs when you cut a chicken into five pieces is to make sure the oysters (still attached via the skin) remain with the leg portions. Follow these steps:

1. Lay the chicken on its back and cut through the skin where the thigh joins the body.

2. Pop the bone from its hip joint and then cut through the socket, but not through the skin on the back (see Figure 5-9).

By stopping the cut just past the socket, you don't sever the oyster.

Figure 5-9:
Removing the thigh.

Illustration by Wiley, Composition Services Graphics

3. **Turn the bird on its side and finish cutting by removing the leg and the oyster in one piece.**

4. **Repeat Steps 1-3 to remove the other leg.**

5. **With both legs removed, cut out the spine.**

   Refer to the earlier section "Cutting out the spine" and Figure 5-5 for instructions on how to remove the spine.

Sectioning the wing portions

As I mention earlier, the beauty of the five-piece cut is that it leaves a nice bit of breast meat on the wings. Follow these steps to cut the wing-breast portions, as shown in Figure 5-10:

1. **Turn the whole breast, with wings attached, skin side up, and press down firmly on the breast to crack the bone (Figure 5-10a).**

   Doing so produces a flat piece of meat.

2. **Sever the first wing portion by cutting through the skin and flesh at the points where each collarbone meets the breastbone (Figure 5-10b).**

   So that some of the breast meat is included in each wing portion, make the cut diagonally at a 90-degree angle, starting from the tip of the shoulder down the breast.

3. **Repeat Step 2 to remove the second wing portion (Figure 5-10c).**

   You're left with the bone-in breast portion.

Figure 5-10: Sectioning the wing portions.

Illustration by Wiley, Composition Services Graphics

Making Boneless, Skinless Chicken Pieces

Boneless, skinless chicken parts are a familiar form of poultry preparation. Lean and clean, these loved cuts are sometimes a mystery in the culinary world. Why would anyone want to remove that crispy skin and those tasty bones? But for many reasons, we do. In this section, I tell you how to produce two boneless, skinless breasts, legs, and thighs each. (You can save the bones for later!)

Removing the skin

When you skin a chicken, you do so in one whole piece. Follow these steps, shown in Figure 5-11:

1. **Remove the head and feet, if your chicken still has them.**

   Refer to the earlier section "Removing the head and feet" for instructions. As always, set the head aside for the stock.

2. **To begin skinning the bird, grab the bird by the wing and use your knife to score around the arm, directly under the arm pit and up and around the top of the wing (Figure 5-11a).**

   When you score around the wings, cut through the skin only, not the meat. By scoring around the wings in advance, you'll be able to peel the skin off easily, keeping it intact.

3. **Slice down the middle of the back, through the skin only (Figure 5-11b).**

4. **Push your fingers under the skin to loosen and free it from the meat (Figure 5-11c).**

   Focus your finger work at the top of the bird, near the neck, breast, and around the wing area.

5. **Here's the fun part: Lift the chicken off the cutting board by its loosened skin and let it hang down (Figure 5-11d).**

6. **Give the skin a few hard, downward shakes, letting gravity pull the skin down and off the legs.**

   This maneuver is similar to the one you use to get your toddler out of his pants, although perhaps not quite as gentle. With a few quick jerks, you should be ready to cut the skin free.

7. **Cut the skin off at the bottom the legs (Figure 5-11e).**

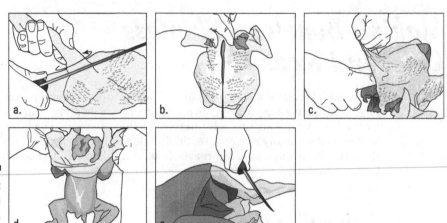

Figure 5-11:
Skinning a
chicken.

Cutting up the skinned chicken

The next step in your endeavors to produce boneless, skinless chicken pieces is to cut the chicken into pieces. To prepare the chicken for boning, you need to do the following (the earlier section "Pieces of Eight: Cutting Up a Fryer" has the detailed instructions and illustrations):

1. **Remove the wings and the legs.**

2. **Cut out the spine and remove the wishbone.**

3. **Split the breast.**

With these cuts complete, you're ready to begin deboning.

Deboning the breast

After you split the breast, you'll find a few rib bones and collar bones left in each one. To get a truly boneless chicken breast, you need to remove these bones. Follow these steps, shown in Figure 5-12:

1. **Lay the chicken breast down on your work surface, rib bones facing away from your body (Figure 5-12a).**

2. **Holding your knife at a slight 45-degree angle and starting just under the ribs on one side, push the blade of your knife down and away from you, pressing up against the bones as you slice (Figure 5-12b).**

   This maneuver is similar to one you'd use to fillet a fish. The ideas is to keep the pressure of the knife blade against the bone as you cut so that you can easily slice the small section of ribs off in one piece, keeping the breast meat on the breast.

3. **Finish your cut and completely remove the rib section from the top of the underside of the breast (Figure 5-12c).**

4. **To remove the collar bone, use your fingers to find where the bone lies in the top of the breast.**

   It may already be poking out a little. Use your hands to pull it out as much as you can to get a better view

5. **Use the tip of your knife to score around the collar bone while the other hand pulls the meat aside, freeing it from the bone to give you a whole boneless breast (Figure 5-12d).**

   Take care not to cut through any valuable breast meat.

6. **Slice down the middle to split the single breast into two, individual breasts (Figure 5-12e).**

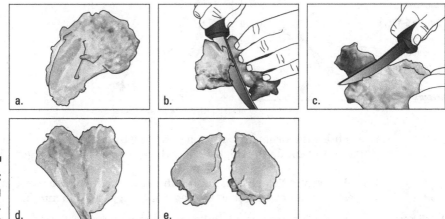

Figure 5-12: Deboning the breast.

Illustration by Wiley, Composition Services Graphics

Deboning the thigh and drumsticks

When you debone the leg, you do it in sections: First, you debone the thigh and then cut the thigh meat away from the leg. Next, you debone the drumstick portion of the leg.

Follow these steps to debone the thigh portion of the leg, shown in Figure 5-13:

1. **With the tip of your knife, score around the bone, pushing the meat away with your fingers and working it free of the bone (Figure 5-13a).**

2. **Continue slicing around the bone in this manner and work the meat under the bone free with your fingers. Then disconnect the meat from the top of the thigh bone completely (Figure 5-13b).**

3. **Remove the entire thigh bone by cutting it free of the knee joint (Figure 5-13c).**

4. **Cut the boneless thigh meat off the leg (Figure 5-13d).**

5. **Repeat Steps 1 through 4 on the other leg.**

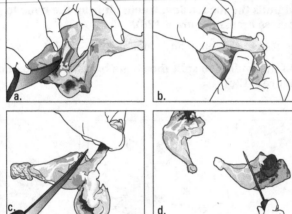

Figure 5-13:
Deboning the thigh portion of the leg.

Illustration by Wiley, Composition Services Graphics

You debone the drumstick portion of the leg by using the same techniques you used to debone the thigh. Follow these steps, shown in Figure 5-14:

1. **Slice around the bone, pulling the meat aside as you work until you can cut the meat free of the leg bone (Figure 5-14a and b).**

   Use the leg bone as a handle to secure your grip.

2. **When all the meat has been removed from the bone, make your final cut (Figure 5-14c).**

Figure 5-14:
Deboning a chicken leg.

Illustration by Wiley, Composition Services Graphics

Impressing Your Neighbors: Boneless Chicken Halves

In this preparation, which I have used many times in my restaurant, you get to further develop your deboning skills. This cut is easy, unique, and impressive, and lends itself to many types of different preparations.

This technique yields two boneless, skin-on chicken halves. Marvelous for stuffing and rolling, cooking under a brick (al matone, or brick-style), marinating and grilling, or pan searing, this cut is a fun way to cut and serve a chicken for two. Each portion has a breast and delicious, dark leg and thigh meat.

To get boneless chicken halves, follow these steps, shown in Figure 5-15:

1. **Remove the head and feet (if necessary) and the wings, following the instructions in the earlier sections.**

   Now you're ready to start the real work of deboning the chicken halves.

2. **Using the tip of your knife, slice against the sternum (Figure 5-15a) and make small slices against the bone, using your free hand to pull the breast meat off (Figure 5-15b).**

   Essentially you're fileting the breast.

   To find the sternum, feel along the top of the bird for a bone that runs down the front of the bird. The breasts are on either side of this bone.

3. **When the meat is free from the top to the bottom of the breast, filet the breast open, removing it from the rib cage but leaving it attached to the leg meat via the skin.**

4. **Open the leg joints by slicing through the hip joint (Figure 5-15c) and cutting the half free of the carcass (Figure 5-15d).**

5. **Turn the bird and repeat Steps 2 through 4 on the other side.**

   At this point, you have a bone-in leg attached to a boneless breast. Now, all you have left to do is to remove the leg and thigh bones.

6. **Turn the chicken half so that the breast is toward you and the leg is facing away.**

7. **With the tip of your boning knife, score around the thigh bone until you can run your fingers along the sides of the bone and grab under it (Figure 5-15e); then pull up with both of your fingers (or knife) at the end of the thigh bone to disconnect the tip of the thigh bone from the meat.**

   Don't cut all the way through the thigh meat. You're simply working around the bone to free it for removal from the thigh, leaving the meat intact. Refer to the earlier section "Deboning the thigh and drumsticks" for instructions on how to score around the thigh bone.

Now that the top of the thigh bone is free, you can use it as a handle to make easy work of deboning the remainder of the leg meat.

8. **Grab the top of the thigh bone with your free hand and use the knife to cut around the bone, systematically cutting into the meat in this order: top, right side, under, left side (Figure 5-15f).**

Continue making these small cuts until you have cut past the joint where the thigh bone is attached to the leg bone. Assist the process by pulling the leg meat down off the bone to see where the ligaments are still attaching the meat to the bone. Keep cutting those ligaments free.

9. **Put your knife down and grab the leg meat; then pull the meat down and off the bone (Figure 5-15g).**

10. **Set the meat and leg bones down on the cutting board and cut through the skin right below the bottom of the leg bone to finish the cut (Figure 5-15h).**

11. **Turn the boneless leg meat right side out by pushing the inside of the meat back into the skin with two hands (Figure 5-15i).**

This move is similar to turning your socks inside out.

12. **Repeat Steps 5 through 11 on the other side.**

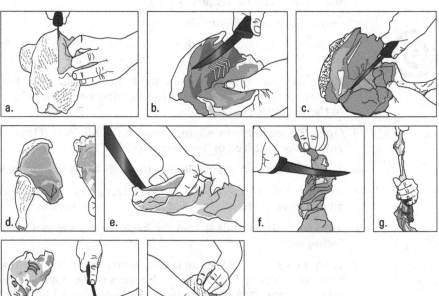

Figure 5-15: Deboning a chicken half.

Illustration by Wiley, Composition Services Graphics

TIP

An easy guide to trussing poultry

Trussing a chicken is important when you plan to cook using any method that requires turning (think, rotisserie) and when presentation is a factor. Trussing pulls the legs and wings close into the bird, creating a compactness that promotes even roasting. Without it, the legs and wings can overcook or even burn before the rest of the bird is done.

You don't need much to truss a bird, only some butcher's twine (cut a piece 3 to 4 times longer than your bird); kitchen shears, scissors, or a sharp knife for cutting the twine; and a cutting board.

You can truss a bird in many ways, but the technique I offer here is quick and easy:

1. Place the bird on your work surface, breast side up, legs toward you. Center the twine under the legs and tail and then cross the free ends over top of legs, making sure the legs and tail are within the loop created. (Think first step in tying your shoes.)

2. Pull the twine tight and flip the bird so that its back is up, its legs are away from you, and its neck is facing you.

3. Pull the twine from the legs up the thighs and along the side of the bird, catching the wing below the shoulder joint and across the wing tips, and pull tightly. The wings will form a V, and the twine should be across the center portion of the V.

4. Cinch and knot the twine at the neck (use a shoe lace knot, which is easy to untie).

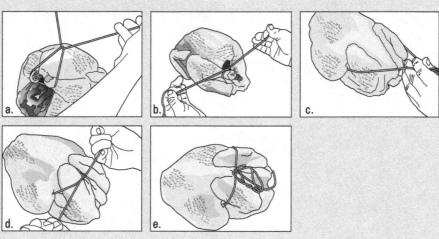

Illustration by Wiley, Composition Services Graphics

Chapter 6

What's Up, Doc? Rascally Rabbits!

Rabbit meat, often overlooked and not given the recognition it deserves, is nutritious, high in protein, and lower in fat than beef, pork, turkey, or even chicken. It's always in season, fed a vegetarian diet, and typically raised above ground, producing a clean, productive, domesticated livestock that thrives in either small- or large-scale farming operations. Rabbits are also sustainable, producing six times more meat with the same amount of water and feed required to produce one pound of beef. That's one efficient lagomorpha — low cost to the farmer and the earth, with a big payoff in quality and profit. In addition to being good meat, rabbit has a mild and delicate flavor.

The body structure of a rabbit is different than that of a bi-pedal animal. A rabbit is a quadruped and more similar to a lamb than to a chicken. For that reason, butchering a rabbit is slightly harder than butchering a chicken (which I cover in Chapter 5), which makes it a great second project for the beginning home butcher. Butchering a rabbit will deepen your familiarity with cuts as you work your way around bones, bones, and more bones.

Cutting Up Fryers and Roasters

Rabbit is an all-white-meat critter. It's delicate, mild, and lean, and is great for stewing, roasting, searing, grilling — you name it. Rabbits, like chickens, are sold whole as either fryers or as roasters:

✔ **Fryers:** Fryers are young rabbits, typically around three months old and weighing between 1½–3½ pounds. Fryer meat is more delicate and has a light pink color. As the names suggest, a fryer is great for a quick, hot heat.

✔ **Roasters:** Roasters are mature rabbits. They're at least eight months old and weigh over 4 pounds. The meat from roasters can be tougher, slightly darker in color, and have a bit more fat. A roaster benefits from a longer, slower preparation.

In this section, I explain how to cut a whole rabbit into eight pieces: two shoulders, two legs, two loin chops, and two bone-in racks. You'll also end up along with some trim, offal, and, of course, bones for the stock pot.

Whether you're tackling a fryer or a roaster, remember that a rabbit carcass may at times be a little tricky to work with because its lean body can be hard to stabilize. On the other hand, the flopping on the cutting board can actually make the rabbit an easy animal to cut because the bones are thin and easy to see. You should be able to make quick work of the eight-piece rabbit technique.

When butchering and deboning rabbit, use a small (5 inches or less), straight boning knife. And if progress is slow (and it may be the first time), remember to return the rabbit to the refrigerator every so often (cover it with a towel or wrap so that the flesh doesn't dry out). This keeps your meat in primo condition while you are working on it.

Removing the offal and silver skin

The first step in processing a rabbit is to remove the *offal*, the internal organs. Rabbits come with the kidneys, heart, and liver still attached. You can easily locate these organs by looking inside the carcass. Regardless of how you choose to cut the rabbit carcass, always begin by taking out the organs. You also need to remove the silver skin, the opalescent connective tissue attached to various pieces of meat.

Follow these directions:

1. **With your hand, pluck out the heart, kidneys, and liver; set aside (see Figure 6-1).**

   You shouldn't need a knife for this task because the carcass will already be split down the center. After the organs are removed, you can do some clean-up knife work by trimming away any blood or connective tissue where the organs were attached.

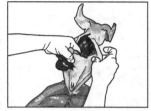

Figure 6-1:
Removing
the offal.

*Illustration by Wiley,
Composition Services Graphics*

You can use these parts in a sauce or pâté; they add a lot of rich flavor and depth to your dish. If you don't plan to use them right away, freeze them for later use.

2. **Look over the carcass and trim any silver skin and sinew from the outside of the rabbit.**

Rabbits can have quite a bit of silver skin, but it's thin, and you can easily pull it away from the body with your hands. Use small slices with your trusty boning knife to peel away sections that don't come off with a gentle tug.

Removing the back legs

After you remove the offal, you're ready to remove the hind legs. Removing a rabbit's hind legs is very similar to removing the legs on a chicken; if you're comfortable with cutting up a chicken, this process should be a piece of cake for you. Follow these steps and refer to Figure 6-2 as necessary:

1. **With the rabbit belly side up and its legs closest to you, take one of the back legs in your hand and make a slice at the base of the leg where the joint connects the leg to the pelvis, the lower half of the spine (Figure 6-2a).**

2. **With one hand on the rabbit loin and the other holding the leg, use your hand to turn the leg out (in a direction opposite the natural rotation) and pop the joint out of its socket (Figure 6-2b).**

3. **Cut the leg free from the body by making a straight slice through the meat, perpendicular to the femur bone, at the base of the leg, just past where you disjointed it (Figure 6-2c).**

4. **Repeat Steps 1 through 3 on the other side of the carcass to remove the other leg (Figure 6-2d).**

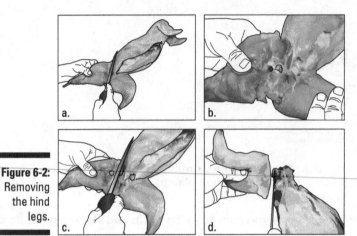

Illustration by Wiley, Composition Services Graphics

Figure 6-2: Removing the hind legs.

Removing the front legs

Taking the arms, or front legs, off a rabbit is the easiest cut in the eight-piece preparation. Because no bone attaches the arm to the body, slicing the arm off is as easy as 1-2-3. You'll feel like a pro making this cut (see Figure 6-3):

1. **Starting at the center of the rib cage, slice down the ribs toward the arm, removing the breast meat from the ribs (Figure 6-3a).**

2. **Continue slicing and pulling the meat away from the ribs until you reach the bottom of the arm; then simply slice the arm free (Figure 6-3b and c).**

3. **Some meat may be on the front leg that is a bit blown out, showing spots of blood, sinew, or fat. Trim this off and add it to the bone and stock pile.**

Cutting through the ribs

At this stage of the rabbit butchery, you've removed the legs and are ready to start working on the loin. The first task is to cut through the rib cage so that you can open up the chest and see inside the rabbit carcass. Then you need to remove the rabbit's diaphragm. Follow these steps (shown in Figure 6-4):

Figure 6-3:
Removing
the back
legs.

Illustration by Wiley, Composition Services Graphics

1. **Stabilize the rabbit carcass with your free hand and slice up through the ribs to split the rib cage in half down the center (Figure 6-4a).**

   Be careful! Know where your knife and fingers are at all times. If you don't feel comfortable making this cut with a knife, use kitchen shears instead. Always trust your instincts: If it feels dangerous to you, find another way. Also, once cut, rabbit ribs are very sharp and can pierce your flesh easily, so handle with care!

   Now that you have the rib cage opened, you can see inside the chest cavity (Figure 6-4b). Inside is a small amount of trim called the *rabbit skirt steak* (or diaphragm), which lies between the abdomen and chest cavity. You may also find some blood and sinew where the heart was attached.

2. **Using the tip of your boning knife, free the rabbit skirt steak and any sinew (Figure 6-4c) and set them aside.**

Figure 6-4:
Opening and
cleaning out
the chest
cavity.

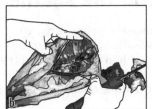
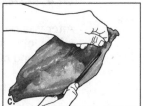

Illustration by Wiley, Composition Services Graphics

Removing the pelvis

The next cut you are going to make removes the remaining pelvis bones from the loin, giving you two clean, professional-looking loin chops. To remove the pelvis, follow these steps, shown in Figure 6-5:

1. **Identify the last vertebra at the base of the loin and slice into the space between the vertebra, cutting down about 1/4 inch (Figure 6-5a).**

   You're not trying to cut all the way through the bone but just down far enough to allow you to split the spine open with your hands.

2. **Take the base of the loin and the pelvis in your hands and bend it backward until you hear the bone crack (Figure 6-5b).**

   This move breaks open the spine and makes cutting the pelvis bone off much easier (and safer).

3. **Cut the pelvis free from the loin by slicing down past the bone, through the remaining meat to remove the piece completely (Figure 6-5c).**

   Add this section of bone to your growing stock pile. You can roast the bones later for a flavorful stock.

Figure 6-5:
Removing the pelvis.

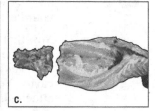

a.   b.   c.

Illustration by Wiley, Composition Services Graphics

Sectioning the saddle

After you remove the pelvis, what you have left is the *saddle,* which is the whole loin, undivided. After you separate the saddle in two, you are left with the rib end of the loin (the *rack)* and the hip end (the *loin).*

To remove the saddle, follow these steps, shown in Figure 6-6:

1. **With the saddle in front of you on a cutting board, identify the tenth rib, which is the last one (Figure 6-6a).**

   You section the rack from the loin at the end of this rib. You can clearly see the tenderloin on the inside, running along both sides of the spine.

Notice where the muscles change from the loin to the rack. Use this as an indicator of where to look for the rib and make your cut.

2. **With the tip of your knife, poke through the flesh; then, using the rib to guide your cut, cut up and out through the flank meat (Figure 6-6b and c). Repeat on the other side.**

   With both sides of the ribs and flank free, you'll follow the same line to cut through the vertebra into the spine.

3. **With your hands on either side of the loin and the cut line in the center, turn the spine backward and crack it to break through the bone, opening up the spine for your final cut (Figure 6-6d).**

4. **Cut through the remaining meat and through the center of the cracked spine to divide the saddle in two (Figure 6-6e and f).**

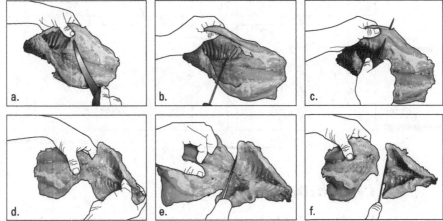

Figure 6-6: Sectioning the saddle.

Illustration by Wiley, Composition Services Graphics

Portioning the loin

Portioning the loin section into two loin chops produces a couple of nice, equal-sized portions to add to your eight piece fryer. Follow these steps, shown in Figure 6-7:

1. **To trim the flank (or belly) off of either side of the loin where the tenderloin muscle ends, make straight cuts parallel to the loin to remove the pieces completely on both sides (Figure 6-7a).**

2. **To divide the loin down the center into two chops, cut through the meat until you hit bone (Figure 6-7b).**

3. **Using the tip of your knife, locate the vertebra to find your entry point; then cut between the discs and down through the center of the spine (Figure 6-7c).**

4. **Take each side of the loin in either hand and bend it backward to crack the bone (Figure 6-7d).**

5. **Cut down through the spine and the remaining meat to divide the loin into two lovely chops (Figure 6-7e).**

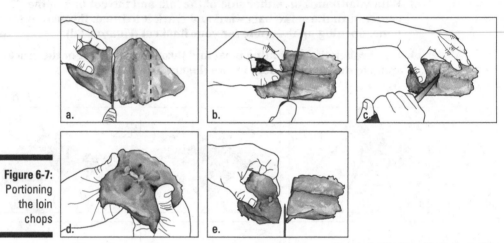

Figure 6-7:
Portioning
the loin
chops

Illustration by Wiley, Composition Services Graphics

Finishing up the rack

Your final task is to separate the rack in two. To complete this task you remove the *chine* (the spine), cut the rack into two pieces, and trim off some of the rib bones. Follow these steps, shown in Figure 6-8:

1. **Using the tip of your knife, cut through the ribs where they connect to the spine (Figure 6-8a); do this on both sides of the chine.**

   This maneuver is called *chining,* and it means removing the spine. If left in, the chine makes cutting through chops more difficult, which is why you want to remove it.

   The bones on a rabbit carcass are thin and easy to cut through, but be careful not to use too much force and mistakenly cut through the bone and into the flesh. If you're having a hard time cutting through the bone alone, rock the tip of the knife slightly back and forth as you move down either side of the spine.

2. **When the chine is free on both sides, free it from the meat by grabbing it with your fingers and pulling it up and out from between the**

racks, continuing to slice under it with the tip of your knife to cut it away from the meat (Figure 6-8b).

3. Cut the racks into two pieces by slicing through the meat in the center of the empty space where you just removed the chine (Figure 6-8c).

4. Trim the spareribs off each of the racks by cutting straight through the ribs about 1 inch away from the base of the rib bones (Figure 6-8d). Add the spareribs to your stock pile.

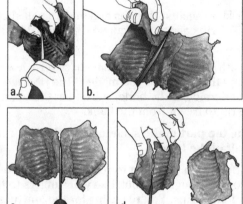

Figure 6-8:
Dividing the
rack.

Illustration by Wiley, Composition Services Graphics

Deboning the Rabbit

A boneless rabbit carcass is the perfect vehicle for stuffing and roasting, making the most of the animal's thin, lean meat. You can stuff the deboned rabbit with forcemeats or perhaps a mixture of bacon, herbs, greens, and Parmesan cheese. After stuffing, roll and tie the rabbit into a cylinder-shaped roast for a perfect preparation that's sure to impress.

Deboning a rabbit requires intermediate skill with a knife. It may take a couple tries to get the process right, which means, of course, that you get to eat more rabbit while you learn — definitely a win-win in my book!

Here's a technique that makes deboning a little easier for non-experts: Always push against the bone as you make small slices with the tip of your knife. Here's why: As the skeleton changes and the bones curve in different directions, accidentally cutting through a muscle is easy, and you don't want to do that. By keeping your knife flush against the bone, you ensure that your knife is always on the right track. Because you're removing only the bones, the meat should stay intact.

Removing the rib cage

The easiest method of deboning a rabbit is to remove the spine and rib cage before you remove the arm and leg bones. Doing so allows you to get most of the tricky deboning out of the way first. When removing the rib cage, you free it from the rabbit one side at a time, beginning at the top of the rib cage and moving down past the loin. Follow these steps, shown in Figure 6-9.

1. **Lay the rabbit on its back with the ribs closest to you. Slice down one side of the rib cage, going from the top down (Figure 6-9a).**

   Be sure to keep your knife against the ribs and make small slices, pulling the meat away from the rib cage as you cut so that the rabbit carcass begins to open up.

   Feel the flank meat with your fingers. You may discover a few ribs that weren't cut free during the first pass you made at the rib cage. If you do, simply slice up the underside of the bone, starting at the tip, and slide your knife down the rib to free it from the flank.

2. **After you release the top part of the skeleton on this side, continue down the bottom half of the loin.**

   Pay particular attention when you start to reach the bottom of the ribs that you don't cut down through the loin. Use your fingers to feel where the ribs end and the loin begins. You're feeling for the loin meat on the underside of the rabbit on either side of the spine. To make sure you don't cut down through the flesh, keep your knife against the bone and slice up and under the ribs.

3. **Slice down past the ribs on the inside of the tenderloin (the inside is the side that butts up against the spine) until you reach the pelvis (Figure 6-9b).**

4. **Use your fingers to start pulling the tenderloin away from the thin section of bone between the tenderloin and the top loin.**

   Pulling the tenderloin back from the spine with your fingers helps you feel where the thin section of bone begins and ends. Use this for reference later when you remove the spine and rib cage completely.

5. **Make a straight cut following the same cut line (close and parallel to the side of the spine you're working on) into the leg to expose the joint; then grab the leg and pop the joint out of its socket (Figure 6-9c and d).**

Essentially what you've done is to loosen the skeleton on one side so that you may remove it in one whole piece after you free it from the other side.

6. **Repeat Steps 1 through 5 on the other side of the rabbit carcass.**

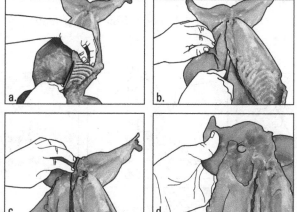

Figure 6-9:
Removing
the rib cage.

Cutting out the skeleton

With the skeleton loosened from both sides of the rabbit, you can now easily cut it free. Follow these steps, as shown in Figure 6-10:

1. **With your free hand, grab the skeleton by the rib cage and lift it up; then use the tip of your knife to slice the upper part of the spine free wherever it is still connected to the flesh. Continue cutting down the spine until you come to the loin (Figure 6-10a).**

2. **Still holding the spine in one hand, take your knife and finish freeing the spine from the lower half of the loin (Figure 6-10b).**

   With both sides of the tenderloin already pulled away from the bone, you just need to run your knife along the underside of the remaining bone, making small slices, and cut the top loin free from the spine.

   Try rolling the spine to one side as you cut the top loin free. Doing so makes seeing where the bone and meat are and aren't connected to each other much easier.

3. **Cut around either side of the pelvis (Figure 6-10c).**

   At this point, the leg bones have already been disjointed, so only the pelvis is still attached to the flesh.

4. **Cut the skeleton completely free of the rabbit carcass by making a simple cut through the remaining flesh attached to the pelvis (Figure 6-10d and e).**

   You now have one partially deboned rabbit with arm and leg bones still intact.

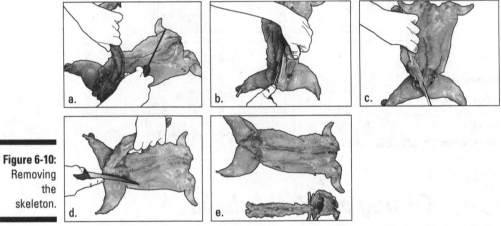

Figure 6-10:
Removing the skeleton.

Illustration by Wiley, Composition Services Graphics

Removing the leg bones

After all the tricky deboning work outlined in the preceding sections, removing the leg bones will feel like a breeze. Follow these steps, as shown in Figure 6-11:

1. **With the semi-boneless rabbit body belly up and legs facing you, score around either side of the femur and leg bones (Figure 6-11a).**

2. **Poke your knife under and out the other side in the center of the femur bone, cut up and out, completely removing the tip of the bone (the ball joint) from the flesh where it was attached to the socket in the pelvis (Figure 6-11b).**

   Be sure to keep you knife against the bone as you cut.

3. **Grab hold of the top of the femur bone, lift it up off the table, and use it to roll the bone away from the meat as you make small slices to remove it completely (Figure 6-11c and d).**

4. **Repeat Steps 1 through 3 to remove the other leg bone.**

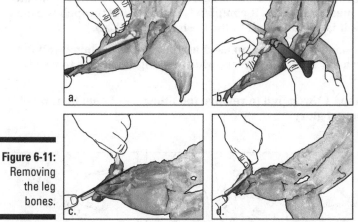

Illustration by Wiley, Composition Services Graphics

Removing the arm bones

The arms are a little trickier than the leg bones to remove because the shoulder blade is attached to each arm bone. For that reason, you use a different technique to debone the arms. Follow these steps, shown in Figure 6-12:

1. **With the rabbit's arms facing you, score around both sides of the arm bones from shoulder blade to tip (Figure 6-12a).**

2. **Run your fingers delicately along either side of the bone to free it from the meat, using your knife as necessary (Figure 6-12b).**

   The meat is thin here, and using your hands along with your knife helps you preserve more of the meat in this area.

   If you ever feel uncertain about where to cut when deboning, take a moment to feel the bones inside the meat with your fingers. Doing so lets you visualize the skeleton and make confident cuts.

3. **Locate the shoulder blade with your fingers and continue cutting up the arm bone until you reach the shoulder blade; then slice directly on top of it (Figure 6-12c).**

4. **Using the tip of your knife, make small slices to score around the shoulder blade. Keep working in this manner until you have cut the flesh free from the shoulder blade.**

   Be sure to keep your knife flush against the shoulder blade as you cut it free.

5. **Using your fingertips, pull the shoulder blade up and away from the shoulder meat; then slice it completely free, cutting through the meat anywhere it is still attached to the flesh (Figure 6-12d).**

6. **Using the same technique you used to remove the leg bone, roll the arm bone away from the carcass while making small cuts to free it (Figure 6-12e).**

7. **Repeat Steps 1 through 6 to remove the other arm bone; then revel in your deboned glory!**

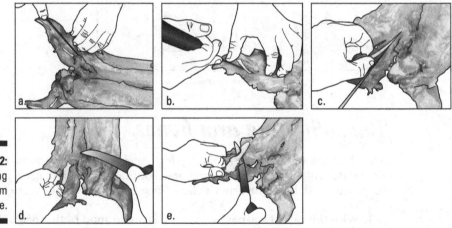

Figure 6-12:
Removing
the arm
bone.

Illustration by Wiley, Composition Services Graphics

Chapter 7

Baaaaack to Basics: Lamb and Goat Butchery

In This Chapter

▶ Discovering the basics of lamb and goat butchery

▶ Identifying lamb primals and subprimals

▶ Making a variety of retail cuts

*I*f you're just learning to butcher, lamb is a great animal to work with. First, it's forgiving. Because many of the subprimals are small enough to be sold or prepared whole or deboned, butterflied and grilled, no matter how you may mess up — and every novice butcher does! — there's always a way to salvage things. Within the loin, rib chops are fairly easy to cut, and the remaining parts need only be minimally processed or cleaned of some fat and set aside for trim. In this chapter, I tell you how to produce lamb or goat primals, subprimals, and retail cuts.

In this chapter, I use lamb for the instructions and illustrations. Keep in mind, though, that any butchery instruction that applies to lamb also applies to goat. If you know how to butcher lamb, you know how to butcher goat, too.

Getting to Know Your Little Bovids

Both lambs and goats are members of the Bovidae family. And they've been popular livestock animals for eons in almost every part of the world, partly because they adapt well to rough terrain that's unsuitable for cows. As a result, these animals have a rich and varied culinary history, showing up as staples in Ethopian stews, as well as French fine dining. They can be enjoyed in a diverse range of cooking preparations: They can be stewed, braised, barbecued, roasted, pan-seared, grilled, ground, and further processed into everything from sausages to *sugos* (sauces or gravies).

The lowdown on lamb

A variety of terms are used for the meat from sheep: *Lamb* refers to a young sheep under one year of age. *Hoggets* are young lambs with two or less permanent incisors (making them one to two years old), and *mutton* refers to older lambs (over two years old), male or female, and is more robust in flavor (some would say gamey). In some countries, *mutton* also refers to domesticated goat meat. Here are some other things to know about lamb:

- ✔ Lamb has a distinct flavor and is a fairly gamey, fatty animal. Typically grown from six to eight months before harvesting, the meat has a rich flavor that often elicits a "love it or hate it" reaction.

- ✔ Lamb is rich in iron and vitamin B, is low in calories, and is rich in *conjugated linoleic acid,* a powerful antioxidant. Despite these benefits, in the United States, the consumption of lamb is pretty low — less than one pound consumed annually per capita.

- ✔ Lamb has four quality grades: Prime, Choice, Good, and Utility. More than 80 percent of domestic lamb produced grades Prime or Choice.

Sheep were domesticated around 10,000 BCE, and to this day, they're still farmed by shepherds who tend to their flocks in the traditional manner. In addition to being used for consumption, sheep are also used for their precious wool, and wool by-products end up in a wide variety of other things, like sausage casings, pet food, fertilizer, cosmetics, shaving cream, lanolin, and even the stuffing in baseballs.

Getting (to know) your goat

Goat has acquired a reputation in the United States for being gamey and tough, but this is simply not the case. Locally farmed young goat is actually quite a delicate, almost sweet, healthy, lean, and sustainable meat. Goat is the most widely consumed meat in the world for those very reasons! If you have the opportunity to support a local goat farmer and purchase a whole animal to butcher at home, take it.

Here are a few other things to know about goats:

- ✔ Goats are efficient animals to farm. They can thrive on land that would starve a cow or horse because they browse (not graze) for their food.

- ✔ Meat goats come in different breeds: Boer, Kiko, Tennessee Meat Goat, Genemaster, Texmaster, Savanna, and Spanish are just a few.

- ✔ Goat is typically harvested anywhere from 3 to 12 months, when the animals weigh an average of 50 to 80 pounds. Depending on the breed, a mature goat can weigh up to 300 pounds.

Covering Lamb and Goat Butchery Basics

Lamb is nearly identical in body structure to goat, with a few differences:

- Lambs are fattier and usually larger than goats, which tend to be much leaner.

- Lambs tend to have a heavier carcass weight than goats. A typical lamb carcass can weigh anywhere from 135 to 160 pounds, depending on where you are purchasing your lamb. The farms I work with produce a much smaller lamb more consistent with a typical goat carcass, which weighs in around 60 or 70 pounds.

- Whole lamb is sold head-off, with the kidneys still attached. Goat is typically sold head-on.

On the bench or on the hook?

Because of their manageable size, lamb and goat carcasses can easily be butchered *on the bench* (your cutting or work surface). That is the method I explain in this chapter.

The other option is butchering *on the rail*, a technique in which the carcass hangs from a hook for butchering. This procedure is the preferred method for butchering beef, because of the weight and unwieldy-ness of the carcass. (I explain on-the-rail butchery in Chapter 12.) If, at some point, you choose the rail over the bench for butchering lamb or goat, keep in mind that the order of steps to breaking down the carcass is different from those in this chapter.

The cuts

Lamb and goat have five primals: shoulder, rack (rib), loin, leg, and breast/foreshank. (In butchery, a *primal* is the section of meat that the carcass is initially cut into. Primal cuts are then cut into *subprimals,* smaller cuts that are then cut further down into retail cuts. For more information on primals, subprimals, and retail cuts, refer to Chapter 3.)

Figure 7-1 illustrates where on the lamb these primals are (remember that this info also applies to goats).

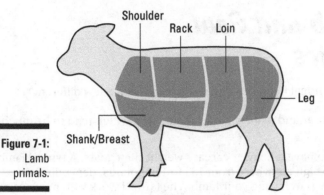

Illustration by Wiley, Composition Services Graphics

Figure 7-1:
Lamb
primals.

Table 7-1 lists primal, subprimal, and retails cuts in a lamb. I also note approximate percentages of the carcass each primal yields. Use this information to get a general idea of how much meat you'll get from each section of the lamb to price out the animal or to plan meals.

Table 7-1	Lamb: Primals, Subprimals, and Retail Cuts	
Primal (%)	*Subprimal*	*Retail Cut*
Shoulder (33%)	Neck, shoulder	Square cut shoulder (whole), Saratoga roast, boneless shoulder roast, blade chop, bone-in arm chop, boneless shoulder chops
Breast (16%)	Breast, foreshank	Riblets, rolled breast, foreshanks, spareribs, Denver rib
Rib (8%)	Rack of lamb, riblets	Crown roast, rib roast, rib chops, boneless rib roast
Loin (12%)	Tenderloin, top loin	Loin chop, double loin chop, tenderloin, loin roast (bone-in or boneless), boneless loin strip
Leg (31%)	Sirloin, leg, hindshank	Whole leg, short cut leg (sirloin off), shank portion roast, center leg roast (shank removed), center slice, American-style roast, boneless leg roast, hindshank, sirloin chop, boneless sirloin roast, top round

As Table 7-1 shows, you get a variety of retail cuts from a lamb (see Figure 7-2). Lamb retail cuts are perfectly tailored for the center-of-the-plate preparations.

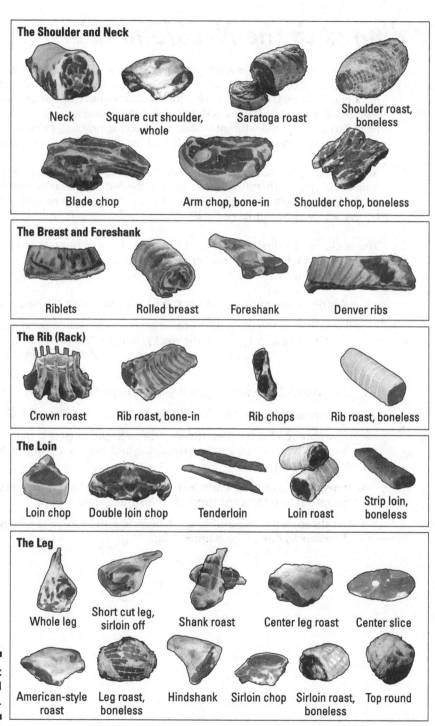

The Shoulder and Neck

Neck   Square cut shoulder, whole   Saratoga roast   Shoulder roast, boneless

Blade chop   Arm chop, bone-in   Shoulder chop, boneless

The Breast and Foreshank

Riblets   Rolled breast   Foreshank   Denver ribs

The Rib (Rack)

Crown roast   Rib roast, bone-in   Rib chops   Rib roast, boneless

The Loin

Loin chop   Double loin chop   Tenderloin   Loin roast   Strip loin, boneless

The Leg

Whole leg   Short cut leg, sirloin off   Shank roast   Center leg roast   Center slice

American-style roast   Leg roast, boneless   Hindshank   Sirloin chop   Sirloin roast, boneless   Top round

Figure 7-2:
Lamb retail cuts.

Illustration by Wiley, Composition Services Graphics

Dealing with the Neck/Shoulder

Although the lamb neck is sometimes not directly named in the listed primals, it's always left on the lamb and is technically part of the shoulder primal. The neck is a delicious, rich cut, with a lot of interconnective tissue that adds flavor to the meat as it breaks down under slow, moist cooking methods. The neck can be deboned, stuffed, rolled and tied for braising or roasting, or cut cross-wise into two or three portions and prepared bone-in like osso bucco.

At my butcher shop, Avedano's in San Francisco, the lamb neck is a prized house favorite. With only one per animal, the butcher usually gets to take it home, but our customers have learned to call ahead and reserve it weekly as we get whole lambs into butcher.

The neck, located just above the shoulder, is easy to remove. You remove it at the point where the shoulder curves into the neck. Follow these steps (shown in Figure 7-3):

1. **Identify your cutting point at the base of the neck (Figure 7-3a).**

   If you look at the grain of the meat on the neck, you'll see a natural line (parallel to the cut edge of the neck) that you can follow to remove the neck in one whole piece. You will be cutting with the grain through the muscle.

2. **Slice through the muscle until you hit bone (Figure 7-3b).**

3. **Cut through the meat around the front and back of the neck (Figure 7-3c).**

   Now that you have cut all the way through the muscle around the neck vertebra, you can clearly see the spine and use the tip of the knife to find an entry point.

4. **Work the tip of your knife into the space in between the vertebra and cut into it and then cut through the spine (Figure 7-3d).**

5. **Cut through the remaining neck muscle and remove the neck from the shoulder in one piece (Figure 7-3e).**

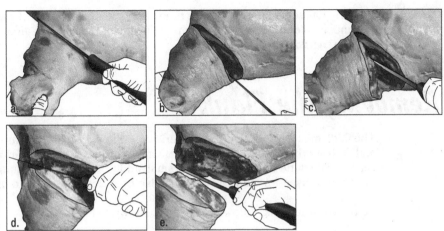

Illustration by Wiley, Composition Services Graphics

Figure 7-3:
Removing
the neck.

Slicing the Skirt Free

The skirt steak, or diaphragm, is located on the inside on the flank. In this section, you remove the two skirt steaks and set them aside for the grind pile.

To slice the skirt free, follow these steps, illustrated in Figure 7-4:

1. **Find the skirt steak inside the saddle.**

   The saddle is the backbone, both loins, and sides on a lamb. The skirt looks like a flap of meat attached to the inside of the saddle (on either side) located where the rib cage ends (away from the spine).

2. **With your free hand, pull up on the top edge of the skirt steak while you use your boning knife to slice near the base of the ribs where the skirt steak is attached (Figure 7-4a).**

3. **Slice down the length of the skirt, from the top to the bottom, removing the skirt steak in one whole piece (Figure 7-4b and c).**

4. **Repeat Steps 2 and 3 on the other skirt.**

Figure 7-4:
Removing
the skirt
steak.

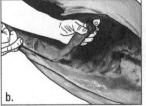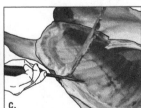

Illustration by Wiley, Composition Services Graphics

To save yourself a little bit of trim time later, spend a moment cleaning any extraneous blood, fat, or sinew from the inside of the rib cage after you remove the skirt steaks.

Removing the Flank

The flank muscles hang off the posterior (rear) portion of the lamb carcass and go from the bottom edge of the rib cage (or breast) to the sirloin end of the leg. The flank on a lamb is fairly small and is usually cleaned and cubed for grind or used for lamb stew. It can also be rolled, stuffed, and braised.

To remove the flank, follow these steps, shown in Figure 7-5:

1. **Identify your cut line.**

   Take a look at the flank muscles and find the ribs. You will see a naturally curved line, starting at the breast, in a half-moon shape that ends at the sirloin end of the leg.

2. **Starting at the breast end of the flank and continuing to the leg, cut the flank free (Figure 7-5a through c).**

   Be careful not to trim the flank too close to the loin muscles, which are close by; otherwise, you may cut into the loin muscle, and that's no good. Cutting into the wrong muscle group means that you damage the cut's full potential (and value). Imagine a tenderloin with an inch shaved off one side — what a waste!

3. **Set the flank aside, under refrigeration, and reserve for grind.**

Figure 7-5:
Removing
the flank.

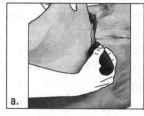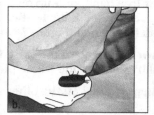

Illustration by Wiley, Composition Services Graphics

Two Tasks in One: Removing the Breast and Foreshank

The breast primal contains the breast and foreshank and is located under the rack and shoulder (refer to Figure 7-1 to see where the primals are in lamb). In this section, you complete two tasks simultaneously — removing the breast and removing the foreshank — in a series of fluid steps.

Here's how it works: You remove the foreshank from the shoulder with a straight cut that runs parallel to the spine, just above the knee. As you continue the shank cutline past the shank and across the ribs/breast, you end up scoring the breast from the posterior edge of the shank to the posterior edge of the rib cage, or breast. After you do one side, you'll do the other side. In the end, you will produce two foreshanks and two lamb breasts. The following sections take you step by step through this process.

Removing the foreshank

Lamb foreshanks are quite meaty and are great braised as an entrée or used for grind, stews, or broths.

To remove the foreshank, follow these steps, shown in Figure 7-6:

1. **Identify your cut line (Figure 7-6a).**

   It starts at the shank, about 1 inch above the elbow joint, and goes in a straight line parallel to the spine to the end of the rib cage. (This is where you just removed the flank.)

2. **Starting from whichever side feels more natural to you, cut through the meat on top of the breast and across the arm, in a straight line, parallel to the spine (Figure 7-6b).**

   Make sure to cut through all the meat surrounding both sides of the arm or foreshank.

   Now that you have cut through all of the meat, you're ready to saw through the arm bone.

3. **Using your bonesaw, saw through the arm bone (Figure 7-6c) and, when you're all the way through, stop.**

4. **Using your knife, cut through the meat on the bottom of the shank to free the arm (Figure 7-6d).**

5. **Wipe the shank clean and set aside, under refrigeration.**

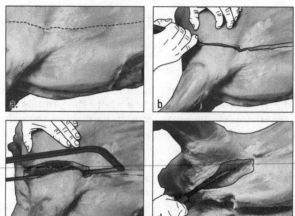

Figure 7-6:
Removing
the fore-
shank.

Illustration by Wiley, Composition Services Graphics

Removing the breast

The breast primal contains the rib tips, which you cut off to produce the rack. When separated from the rest of the breast (or in this case the rack), this small section of ribs is called a Denver rib. Although neither home cooks nor restaurants use the breast extensively, it's delicious stuffed and braised (bone-in or boneless).

Because you've already scored the breast (refer to the preceding section), all that you need to do now is saw through the ribs to remove the breast from the carcass. Follow these steps, shown in Figure 7-7:

1. **Cut through the meat on top of the breast, where the foreshank was just removed (Figure 7-7a); then grab your saw.**

2. **Saw through the ribs and breast plate, down the center of your previously scored cut line (Figure 7-7b and c) until the breast portion is free.**

3. **Repeat Steps 1 and 2 to remove the foreshank and breast portion on the other side of the carcass; then set the lamb breasts aside, under refrigeration, until you are ready to process them into retail cuts.**

   From the lamb breast, you can get the lamb breast riblets, rolled breast, ground lamb, and, of course, a whole breast.

Figure 7-7:
Removing
the breast.

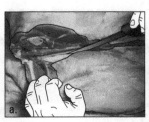 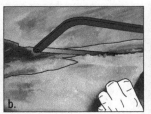 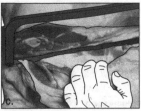

Illustration by Wiley, Composition Services Graphics

Removing the Hindshanks

The hindshanks on a lamb may be bigger and meatier than the foreshanks, but you can remove them just as easily, by either sawing through the bone or by using a boning knife to cut through the knee joint. The following sections explain both methods.

Why choose? You can practice both methods by using one method on the first hindshank and the other method on the second hindshank. This is an especially good way to get a little knife-and-joint practice in.

Using a saw to remove the hindshank

To remove the hindshank by sawing through the bone, follow these steps, illustrated in Figure 7-8:

1. **Cut through the *shank meat,* the meat above the knee (Figure 7-8a).**

   If you need help identifying where to cut, bend the shank back and forth so that you can see where it bends. Cut above that point.

2. **Cut around the back of the shank, making sure to cut through all the meat until you hit bone (Figure 7-8b).**

   After you've cut through all the meat, the only thing left to do is to saw through the bone.

3. **With your bonesaw, saw through the shank bone (Figure 7-8c), stopping as soon as you're through the bone.**

4. **Use a knife to cut through any remaining meat and remove the shank in one piece.**

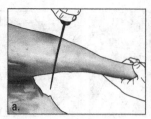 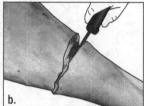 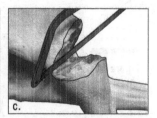

Figure 7-8: Removing the hindshank with a saw.

a. b. c.

Illustration by Wiley, Composition Services Graphics

Using a boning knife to remove the hindshank

To remove the hindshank with a boning knife, follow these steps, shown in Figure 7-9:

1. **Cut through the knee joint, starting at the front of the shank (Figure 7-9a).**

   If you need help identifying where to cut, bend the shank so that you can see where it bends, and cut above that point.

2. **Cut the shank free from the leg (Figure 7-9b).**

Cutting through joints can be a little tricky, because it takes finesse, not brute strength. Use your knife and the joint's natural motion to find the right place to make your cut. And remember that you can't cut through thick bone with a knife; attempting to do so is a way to seriously hurt yourself. If you're pushing really hard, you're doing it wrong.

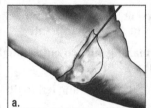 

Figure 7-9: Removing the hindshank with a boning knife.

a. b.

Illustration by Wiley, Composition Services Graphics

Removing the Shoulder

The primal lamb shoulder contains four (sometimes five) rib bones, the arm, and the blade and neck bones. Within the shoulder are many small muscle groups that can make producing chops or steaks a little difficult. But if you persevere, you can produce blade chops, round bone chops, or smaller, butterflied cuts that you can grill or pan sear. The shoulder cuts are also well-suited for tying into roasts, cubing into stew meat, or grinding into lamb burgers or a delightful merguez sausage.

To remove the shoulder, you separate it from the loin between ribs four and five. To accomplish this task, you first have to count to the right rib, cut through it, cut through the shoulder meat, and then finally saw through the bone. Follow these steps, illustrated in Figure 7-10:

1. **Looking on the inside of the carcass and beginning at the first rib (located at the anterior end of the spine), count in four ribs (Figure 7-10a).**

   This puts you at rib five.

2. **Cut through the meat in between the fourth and fifth rib (Figure 7-10b).**

   Keep your knife pressed against the fourth rib when you make your cut and slice down the rib from the top until you reach the *chine* (spine).

3. **Turning your focus to the outside of the carcass, follow the line that you started on the inside and cut through the shoulder muscles with a straight cross-wise cut (Figure 7-10c). Repeat on the other side.**

   Make sure you cut through the shoulder and loin muscles underneath the spine.

   This step cuts all the meat free from the shoulder. Now you simply need to saw through the spine to remove the both shoulders from the rib.

4. **Using your bonesaw, saw through the spine, stopping once you have sawed through (Figure 7-10d).**

5. **With a large breaking knife, cut through the muscles, separating the shoulder from the loin (Figure 7-10e).**

   Now you're left with two individual shoulder primals connected to each other in one whole piece. Time to separate them.

6. **Saw down the middle of the spine to separate the shoulders into two pieces (Figure 7-10f).**

7. **Set the shoulders aside, under refrigeration, until you're ready to process them into retail cuts.**

   The shoulder can produce a lamb shoulder square cut, shoulder blade chops, Saratoga chops, lamb stew meat, ground lamb, tied roasts, round bone chops, or arm chops.

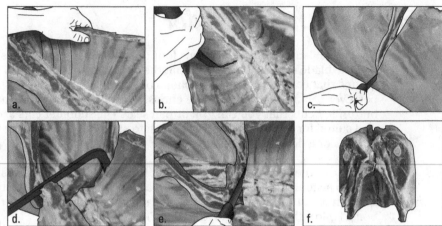

Figure 7-10: Removing the shoulder.

The Leg

The legs are located at the posterior portion of the carcass. You remove them from the loin with a straight cut across the hipbone cartilage or through the backbone vertebra at the base of the spine. At that point, you split the leg primal into two legs and usually debone it.

From the leg, you can produce bone-in and boneless steaks, as well as smaller, tied roasts. You can also tie a whole boneless leg roast it, butterfly it, cube it into stew meat, or use it for ground lamb.

Removing the legs from the loin

After you remove the shanks (refer to the preceding section), you are left with two legs, the rib, and loin. The next cut you make will remove the rib and loin from the two legs.

To remove the legs from the loin, follow these steps, shown in Figure 7-11:

1. **To find your cutting point, open the stomach cavity and look inside the loin for the two tenderloins running on either side of the spine (Figure 7-11a).**

   You remove the legs from the loin, using a straight cut immediately anterior to the hipbone and leaving no less than 1½ lumbar vertebrae on the leg. You section the loin from the leg at the anterior vertebra directly above where the spine curves into the tail.

2. **Run your fingers along the spine close to where it curves into the tail and look closely for each vertebra.**

   You can identify a vertebra by two raised bumps close to each other with a white space between them. This white space is the disc.

3. **Insert your knife into the disc between the vertebra just above the tail (Figure 7-11b).**

   Rock your knife forward and backward to cut further down into the vertebra and through the tenderloin. This prep cut ensures that the loin comes away easily after you complete the next few steps.

   Following these instructions, you are cutting through the tenderloin. If your end goal is to cut porterhouse or T-bone steaks from the loin, this is the correct way. If you want to remove the tenderloin in one piece, you may do so prior to this step.

4. **With your free hand, grab the lower part of the flank or belly (near the leg) and cut down the flank, slicing to separate it from the leg in a reasonably straight line. Stop cutting when you reach the cut line you created between the vertebra in the loin (Figure 7-11c).**

5. **Repeat Step 4 on the other side of the carcass to release the belly from the legs on both sides (Figure 7-11d).**

6. **Following the same line you created between the vertebra, use a large butcher's knife or cimeter to make a straight cut through the loin (Figure 7-11e).**

7. **Break the spine (Figure 7-11f).**

   Put your knives down and grab under the bottom of the loin, lift up with your arm, and "crack" the spine.

8. **Cut through the remaining meat underneath the spine to remove the legs from the loin (Figure 7-11g).**

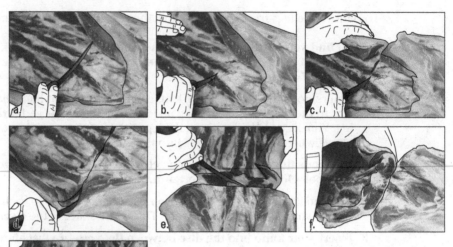

Figure 7-11:
Removing
the legs
from the
loin.

Illustration by Wiley, Composition Services Graphics

Sawing the legs in two

After you remove the legs as I explain in the preceding section, you're not quite done because the legs are still connected to each other. Now you simply need to separate them by sawing down the center of the spine.

To separate the legs, follow these steps, shown in Figure 7-12:

1. **Use your bonesaw to create a groove in the center of the spine (Figure 7-12a).**

   This groove allows you to anchor the saw, thereby ensuring that the saw doesn't slip and create jagged edges in the bone.

2. **Continue sawing until you have passed through all the bone (Figure 7-12b) and then stop.**

   Use your free hand to push the legs open, helping to split the spine apart as you work.

3. **After you have broken through the bone, use your large butcher knife or cimeter to cut the legs in half down the center in a strong, straight line (Figure 7-12c).**

4. **Set the legs aside, under refrigeration, until you're ready to process them into retail cuts.**

   You can seam out the muscles in the legs, tie roasts, cut steaks, or cube into stew meat, kabobs, or grind. You can also debone the whole leg.

Figure 7-12:
Removing
the legs
from the
loin.

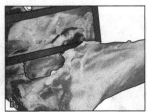

Illustration by Wiley, Composition Services Graphics

Working with the Rib

At this stage in the butchery process, all you have left to work on is the loin and the rib. The task now is to separate the two.

The rib primal has eight ribs on it. The rib is typically cut into smaller individual rib (or lamb) chops. Lamb chops are tender and flavorful and great broiled, grilled, or pan seared.

The famous crown roast is also produced from the rib by tying two racks together in a way that forms a round shape that looks a lot like a crown. You may or may not chose to *french* the lamb rack (cutting the meat away at the tip of the bone to expose the bone), but either way, the rack can be roasted whole and sliced in-between each bone during carving.

Separating the rib from the loin

To separate the rib from the loin, follow these steps, illustrated in Figure 7-13:

1. **Find the last rib and, with the tip of your boning knife, cut between the vertebra near the last rib (Figure 7-13a).**

   When separating the rib from the loin, you leave one rib on the loin; that's why you cut behind it in this step.

2. **Cut down the rib from the spine to the belly behind the last rib (Figure 7-13b).**

3. **Repeat Steps 1 and 2 on the other side (Figure 7-13c).**

4. **Put your knife down and, with one hand underneath and one hand on top, lift up on the loin and crack the spine (Figure 7-13d).**

5. **Cut through the center of the spine with a large butcher knife or cimeter to separate the rib from the loin (Figure 7-13e).**

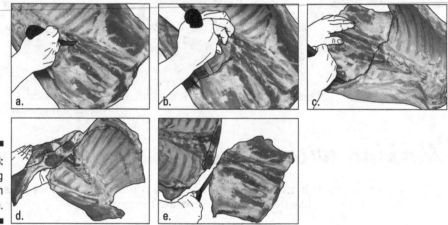

Figure 7-13: Separating the rib from the loin.

Illustration by Wiley, Composition Services Graphics

Chining the rib

After you remove the rib from the loin, you want to remove the spine so that you have two rib sections. The chine, also known as the spine, is a pesky piece of bone when you want to cut chops. But you have a few options: You can separate the chops by cutting in between each vertebra. You can saw through the chine at the point you intend to portion the chops. Or you can remove the chine completely, allowing you to easily cut chops. In the next steps, you remove the chine.

To remove the chine, follow these steps, shown in Figure 7-14:

1. **Lay the rib skin side down on your cutting board, with the spine perpendicular to you.**

2. **On either side of the spine, identify where the spine ends and the ribs attach to the spine.**

3. **With your knife, score a parallel line along each side of the spine (Figure 7-14a).**

   Run your knife down each side several times. This move does two things: It identifies your cut line, and it creates a groove for your bone-saw during the next step.

4. **Place the bonesaw inside the groove on one side of the chine and saw at 45-degree (inward) angle between the rib bones and the spine (Figure 7-14b).**

   Be very careful not to saw past the bone and into the muscle. Keep your eye on the visible end of the spine so that you can gauge how thick the bone is and stop sawing before you reach the meat.

5. **Turn the loin 180 degrees and repeat Steps 2 through 4 on the other side (Figure 7-14c).**

   Before moving on to Step 6, make sure that your saw work has released the spine from the ribs. If not, pick up your saw and finish the task.

6. **With a boning knife, cut down either side of the chine (Figure 7-14d).**

7. **With your free hand, lift up on the spine to pull it up and away from the racks; then cut it free from the rib (Figure 7-14e).**

8. **With a clean towel, wipe the meat clean of bone and bone dust.**

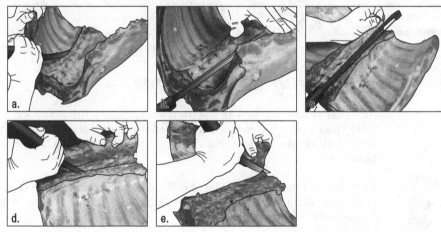

Figure 7-14:
Chining
the rib.

Illustration by Wiley, Composition Services Graphics

Cutting Denver ribs

Now that you have produced two racks, notice that the bones are fairly long and some of the breast plate is still attached. The next task is to trim the top portion of the ribs off, producing Denver ribs. Follow these steps, shown in Figure 7-15:

1. **Determine the amount of ribs you would like to leave on the rib rack.**

   Four inches or so is about right.

2. **With your boning knife or cimeter, score on top of the bone where you want to remove the portion of Denver ribs (Figure 7-15a).**

3. **Saw directly on top of your scoring cut, stopping as soon as you've sawed through the bone (Figure 7-15b).**

4. **Using your knife, cut in between the sawed ribs, removing the Denver ribs from the rib rack (Figure 7-15c).**

Figure 7-15:
Cutting
Denver ribs.

a.   b.   c.

Illustration by Wiley, Composition Services Graphics

Portioning the rib chops

Now that you've finished all the prep work, cutting the rib chops is an easy task. You can cut the chops any desired thickness: single, bone-in chops; double rib chops; or even thin chops alternating between bone-in and boneless.

Regardless of the size, use a large knife and make deliberate cuts, dragging your knife through the meat in one solid motion. Doing so ensures that your chops are straight and attractive.

Before you cut the rib chops, however, you need to remove the small piece of shoulder blade that remains. Simply slice above and below the remaining piece of shoulder blade and use a hand/knife combo to remove it.

Then to cut the chops free from the rack, simply use a butcher knife or cimeter to slice between the rib bones, as shown in Figure 7-16.

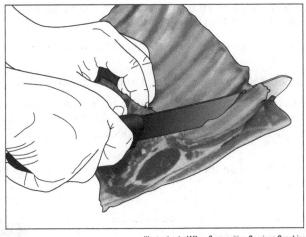

Figure 7-16:
Cutting
chops from
the rack.

Illustration by Wiley, Composition Services Graphics

The Loin

The loin, located between the rib and leg primals, contains one rib (number 13) and portions of the backbone or spine. The loin is best known for tenderloin, top loin steaks (New York strips), and bone-in loin chops or lamb T-bone steaks. Within the loin, you have the loin eye muscle, the tenderloin, and the flank.

Loin meat is prized because of its tenderness and versatility in preparations. For a real treat, prepare it using dry-cooking methods like oven roasting, grilling, pan-searing, or broiling. Other popular preparations of the loin are boneless tenderloin medallions and other boneless chops and roasts.

With all the butchery you've done so far, you're left with a loin that has a portion of the flank still attached to either side. You can prepare the loin in a number of ways: You can leave the flank on the loin, split the loin down the center of the spine, or debone the loin. You can also wrap and tie the flank around the loin and then slice it into elegant chops. Or you can trim off the flank and use it for grind or cut it up for stew meat.

In this section, I tell you how to trim off the lamb flank and cut double loin chops. Follow these steps, illustrated in Figure 7-17.

1. **Cut the flank meat from either side of the loin (Figure 7-17a).**

   Cut about 1 inch away from the tenderloin. It's okay to leave some "tail" on the loin chops.

2. **Cut in between each vertebra and through the meat, portioning your loin chops (Figure 7-17b).**

   Be careful to make straight, cross-wise, even cuts so that you don't end up short at the end.

3. **Using a meat cleaver and a mallet, knock each chop off (Figure 7-17c).**

   Stick the blade of the cleaver in between the vertebra and give the cleaver a good whack with the mallet. For the best results, work with the first 2 to 3 inches of the "point" of the cleaver; you get better leverage this way.

   The cleaver should do the trick, but if you'd rather, you can use a knife to cut the chop free.

Figure 7-17:
Cutting chops from the rack.

a. b. c.

Illustration by Wiley, Composition Services Graphics

Part III
Pork Butchery

"Sausages ready in 6 hours!"

In this part . . .

In this part, I tell you how to butcher a whole hog into cuts that you recognize on your dinner plate. Although cutting larger animals can be challenging, this part gives you the confidence and know-how to take on the task. Work your way up beast by beast and practice developing good habits and knife skills.

Chapter 8

Porky Pig: Understanding the Beast

. . . .

In This Chapter

▶ Getting familiar with pork

▶ Boning up on the pork butchery basics

▶ Distinguishing between pork primals, subprimals, and retail cuts

. . . .

***P**ork,* a culinary term for meat from domesticated pigs, is very versatile meat. It's both lean and fatty, and richly flavored and mild all at once. Although you can enjoy pork fresh and cook it as you would any other meat (grilling, roasting, broiling, and so on), much of the pork produced and consumed in the United States is actually cured or further processed. In fact, 75 percent of the American pork diet comes from processed pork like ham, bacon, and sausage.

Because there is almost no waste on a pig — that is, nearly every piece and part of a hog can be used — pork is an efficient, economical choice for restaurant chefs butchering in their kitchens, and it's a safe bet for adventurous home cooks. In this chapter, I introduce you to pork butchery. Here you can find information about pork consumption, basic principles of working with hogs, and a breakdown of the primals, subprimals, and retails cuts you'll produce in the next chapters.

Pigs are heavy animals, and butchering a pig carcass, whether whole or as a side, is a daunting task. Before you try your hand at butchering pork, make sure you have the right equipment set up and, if it is your first time, have an assistant to help you.

Porkopolis, Harry Truman, and other bits of pig trivia

Here are a few fun and interesting facts about pork and pigs:

✔ Pigs are a member of the Suidae family and include many species of wild pig, like wild boar and warthogs. Domesticated pigs are recorded to have been habituated over 10,000 years ago.

✔ Pigs are natives of the Old World. Hernando de Soto, the "Father of American Pork," is said to have brought 13 pigs with him when in came to America in 1539. Over 470 years later, pork is a standard, domesticated, meat mainstay.

✔ In early America, pioneers traveled across the country carrying small crates of pigs with them to breed upon arrival in their new settlements. The expansion of farming as it moved westward with the settlers gave way to a need for processing in the Midwest. Outside of Buffalo, New York, the country's chief processing center, pigs were first commercially slaughtered circa 1819 in Cincinnati, Ohio. Cincinnati's thriving pork packing industry earned the city the nickname Porkopolis. Not to be outdone,

Chicago became the leading meat-packing center after the Civil War and even inspired the great American poet Carl Sandburg to proclaim it "Hog Butcher for the World" in his 1914 poem "Chicago."

✔ The saying "living high on the hog" was a military term that originated from enlisted army men who received pork shoulder cuts (roasting, braising cuts), while the higher ranking officers received top loin cuts (tender grilling cuts).

✔ According to the Pork Checkoff, a national pork board, the U.S. slaughters a staggering 444,925 head of hog each day, and it ranks number three in the world for pork consumption, coming in under China and Europe at a mere 10,187,000 metric tons a year. We would all benefit from eating quite a bit less meat and purchasing our meat from local butchers who support small farming practices. But clearly the U.S. loves a good piece of pork.

✔ Harry Truman said, "No man should be allowed to be president who does not understand hogs."

Pork and Pigs: Getting to Know the Beast

Pork, the number one most consumed meat, makes up 40 percent of the world's meat consumption. The average American consumes 43 pounds of pork per year.

Although you may think that Americans' love of pork would make the U.S. the world's largest pork producer, that distinction goes to China, whose love affair with the hog dates back to around 4000 BCE when an emperor decreed that all Chinese citizens were to breed and raise hogs. Today, all over the world, pigs are farmed for consumption, sold fresh, cured into hams, smoked for bacon, and processed into sausage and many other products.

Pork production

Although pork is available for purchase year-round, pork production is affected by the season. As a home butcher or a pro, you need to be aware of these factors because this can affect the price of the animal per pound and also affect the fat content. The price of pork may vary due to the seasonal conditions in combination with product demand.

Hogs eat less in hot weather and therefore grow more slowly during the summer months. Also, hogs don't breed as well during the heat, and better breeding conditions arise during the cooler months. Even though pork production takes place year round, always check with your local farmer to see when they have pigs coming up for harvesting.

Weighty matters: Making sense of pork poundage

As a butcher who orders meat for home or business, you need to know several industry terms regarding pricing and the weight of the animal. This information will help you when you are communicating directly with farmers.

Here are the weights you need to know:

- ✓ **Live weight:** *Live weight* is the weight of the living animal.

- ✓ **Hot weight:** *Hot weight* is the weight of the carcass after the pig has been slaughtered and the hide, innards, organs, and (sometimes) the head have been removed.

- ✓ **Chilled weight:** *Chilled weight* refers to the weight of the carcass after it's been refrigerated for awhile. During refrigeration, the carcass loses a bit more weight because of muscle shrinkage and loss of moisture.

An average hog weighs around 260 pounds (live weight) producing a carcass yield of about 200 pounds (hot weight).

Pork's USDA identification categories

Pigs are typically harvested anywhere from six to nine months of age, and although they are omnivores and will forage for their own food, in industrial farming practices, they are fed a diet of corn.

Pork, unlike beef, is not graded. Instead, it is categorized according to one of two USDA identifications: acceptable and utility.

- ✔ **Acceptable pork:** Pork identified as acceptable is used for consumption. The category applies to the whole animal and is determined by the USDA based on its standards and procedures.

- ✔ **Utility pork:** Pork identified as "utility" can be used for pet food and in a fascinating variety of "co-products" that include glue, plastics, crayons, cosmetics, linoleum, fertilizer, floor waxes, garments, dyes, nitroglycerine, cement, and pharmaceuticals.

Fundamentals of Pork Butchery

Understanding primals and subprimals is fairly simple if you allow logic to move you through a carcass. In fact, butchery is a logical series of steps, designed to most efficiently and effectively portion and produce the animal into edible cuts. First, cut the animal into workable pieces (primals), then smaller pieces, keeping musculature in mind (subprimals), then finally into individual portions (retail cuts). As you butcher, you process the hog from a carcass into what you recognize on the dinner plate.

It is sometimes difficult to understand this path because so much conflicting information is out there. The conflict originates from the fact that you are viewing the information from many different vantage points: every country, culture, and individual has a different interpretation of the animal, and for each of them, what they see and do is *the only right way*. To clear up what can be murky meaty waters, having a common language and recognizable terminology is vital to both consumers and butchers. Refer to Chapter 3 for the terms you need to know.

Inspecting the carcass

A hog carcass will typically come in either as a whole animal or in two sides (half of a pig, split down the middle) with or without the head attached.

When the whole animal is delivered, you want to inspect the carcass for quality. The skin should be light in color, not tacky or sticky, and the carcass should not have a strong odor. If the pig comes to you head on, take a look at the eyes to make sure they are not cloudy or sunken in. Also feel free to ask questions of the distributor or farmer: "When was the pig harvested?" "What did it eat?" "How were its living conditions?" All of these variables contribute to the quality of the meat, good or bad, so don't be afraid to advocate for your right to answers.

Paying attention to safety issues

Although it's not as heavy as a cow, pigs are still heavy enough and unwieldy enough that you need to take extra care when butchering them. Before you decide to tackle pork butchery, make sure you observe the following safety guidelines:

- ✔ **Make sure your knives are sharp and that you know how to use them safely.** If you don't, review the safety information in Chapter 4 and delay your foray into pork butchery until you're confident you can do so safely.

- ✔ **Wear the appropriate protective gear.** Chain-mail aprons, cut gloves, scabbards, steel-toed boots, and a slip resistant floor mat can help protect you from an accidental slip of your knife or your footing.

- ✔ **Have a partner nearby to help you.** Pigs are heavy. A whole hog can weigh anywhere from 150–250 pounds (a side of pork may be more manageable for beginners). If you can't handle the weight of the animal yourself, you need a partner who can lend a hand.

- ✔ **Give yourself enough time to do the job properly and safely.** Hurrying through the job is an invitation for an accident. Plan on the task taking several hours and then take your time.

- ✔ **Make sure you have enough space to store the meat both during butchery and then when you're done.** Only Lucille Ball can make running around like a madwoman tucking hunks of meat into closets, drawers, and eventually the basement furnace funny.

Getting Familiar with Pig Primals, Subprimals, and Retail Cuts

One of the main challenges in butchering whole animals, aside from the task itself, is knowing what you are cutting and what you are trying to produce. The information in this section will get you well on your way to recognizing how the animal is broken down, first into primals, then subprimals, and finally into retail cuts.

First and second cuts: Primals and subprimals

Pork has four primals: the shoulder, the loin, the belly, and the ham. After the hog is broken into primals, you can then portion it out into subprimals.

Figure 8-1 shows the pork primals and indicates the subprimals each primal yields.

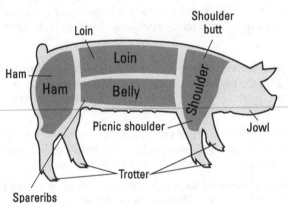

Illustration by Wiley, Composition Services Graphics

Figure 8-1:
Pork primals
and
subprimals.

Table 8-1 lists estimated percentages that each subprimal yields, based on total chilled carcass weight. Use this information to average how much meat you'll be getting from each section of the pig, which can help you with meal planning or, if you're a professional butcher, for pricing out the animal.

Table 8-1	Yield of Pork Subprimals, in Percentages
Subprimal	*Percentage of Carcass (Skin on)*
Pork Boston butt	10%
Pork shoulder, picnic	11%
Loin	25%
Sparerib	5%
Belly	16%
Ham	25%
Jowl	2%
Trotters	2%

Note: The remaining 4 percent is made up of bones and some "unusable" trim such as glands, sinew, and connective tissue.

TIP

The head is not typically included in the primal list, but that doesn't mean it's something you want to throw away. Pork jowls are always appreciated: They can be used for *guanciale* (a fresh Italian bacon) or cured and sometimes smoked. So don't pass up on the hog head!

The retail cuts

Retail cuts are, not surprisingly, those sold in a retail market. As I mentioned in Chapter 2, having a standardized resource for retail cut names is important to creating a language around meat that both consumers and butchers can understand. The North American Meat Processors Association (NAMP) guide and *The Art of Beef Cutting,* by Kari Underly, are good resources.

Cuts from the shoulder

Pork shoulder is reasonably priced, has a lot of great marbling, and is a favorite for barbecue enthusiasts. These cuts are shown in Figure 8-2.

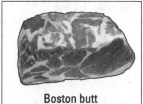
Boston butt

Pork shoulder picnic

Pork shoulder roast

Pork shank

Pork hock

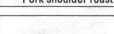
Bone-in country-style ribs

Figure 8-2:
Retail cuts
from the
pork
shoulder.

Pork steaks

Illustration by Wiley, Composition Services Graphics

Cuts from the ham

The ham is the pork leg, a lean muscle system that houses many smaller muscles that can be used for roasts and cutlets and is great for curing (think prosciutto!), brining, and smoking. Figure 8-3 shows the retail cuts you can get from the ham.

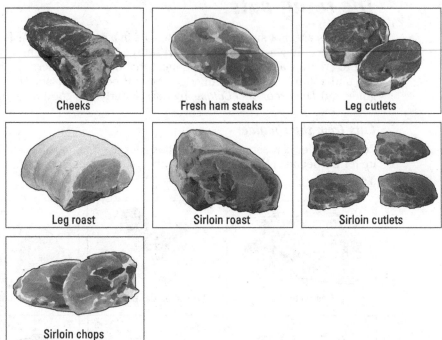

Cheeks

Fresh ham steaks

Leg cutlets

Leg roast

Sirloin roast

Sirloin cutlets

Figure 8-3: Retail cuts from the ham.

Sirloin chops

Illustration by Wiley, Composition Services Graphics

Retail cuts from the loin

The pork loin houses many of the prized center-of-the-plate cuts like bone-in pork chops, baby back ribs, and boneless pork loin. Great for searing and roasting in individual portions, the loin can also produce great roasts. Figure 8-4 shows the retail cuts you can get from the loin.

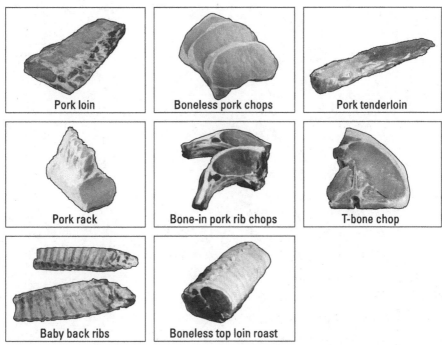

Figure 8-4:
Retail cuts
from the
loin.

Cuts from the belly

What can I say about pork belly? Well *BACON* is one word, but in addition to salt curing and smoking the pork belly, it is fabulous cut into portions and slowly braised and served over grits or polenta. You also find spareribs located on the pork belly. Figure 8-5 shows the retail cuts you can get from the belly, which happens to be one of my favorite cuts.

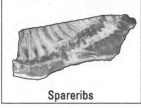
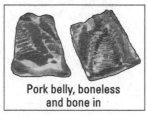

Figure 8-5:
Retail cuts
from the
belly.

Spareribs

Pork belly, boneless
and bone in

Chapter 9

Pork: Cutting It Up

In This Chapter

▶ Cutting a hog into primals

▶ Discovering different sawing techniques

Pork, whether fresh, ground, smoked, or cured, is downright delectable: It's both mild and rich in flavor, lean and fatty, and houses a variety of cuts both small and large. Pretty much every part of this animal can be consumed via an infinite catalog of mouthwatering preparations: grilled, braised, roasted, pickled, stewed, fried, pan seared, and more.

In this chapter, I explain how to cut a whole hog into primals. Butchering a pig carcass produces very little waste — only a handful of glands and bits of cartilage — so it's an easy "risk" to take when furthering your meat cutting skills.

A Bit of Advice before You Begin

In the last decade, pork has come back from its exile in the land of bland, dry, "other white meat" classification. Pork is actually red meat, and chefs have begun insisting that it be treated once again as a full-flavored meat. Second in line only to beef per our meat consumption, pork has also seen a rise in heritage breed farming, which strives to produce richer meat, better fat, and more complex flavors. Finally, fat gets the respect it deserves!

As you begin your adventure in porcine butchery, keep in mind that a hog is typically 9–12 months old when it is slaughtered and can weigh approximately 220–250 pounds. So butchering one is a pretty big task. Here are a couple of suggestions if you're not sure you're ready to tackle one on your own:

✔ Get a couple friends together and purchase a hog from a local farm. Then share the cost (and the meat) and work together to break the pig down into portions.

✔ Consider buying a side of pork rather than a whole hog.

✔ Practice on a suckling pig. Some of the cuts won't be applicable, but it can help you envision the task "in miniature."

Removing the Head

Depending on where you sourced your pork, it may come to you with the head still attached. If you receive a whole hog as two sides of pork, the head may be attached to one side only, or it may be removed completely. (Of course, if your hog came sans head, skip to the next set of instructions.)

To remove the head, you need to identify your cut line, slice through the neck or collar with your boning knife, saw through the bone, using your bone-saw, and make the final cut with your boning knife. Follow these steps for a successful start to pork butchery (see Figure 9-1):

1. **Find your cut line between the shoulder and neck (Figure 9-1a).**

   Take a look at the head and the shoulders. Feel the shoulder and identify the bones inside of it. After you find the edge of the shoulder/arm bone, identify a two- to three-fingers width in front of the bone (going toward the head) as your cutting line.

   You want to leave as much valuable shoulder meat with the rest of the shoulder as possible, so make sure you don't cut too close to the shoulder. Erring your cut more toward the head is the better choice.

2. **Holding your boning knife in a dagger grip (also called a *butcher's grip;* see Chapter 4) and following the cut line you identified in the Step 1, cut down the neck as straight as possible from the top (the spine) to the bottom (under the chin) (Figure 9-1b).**

   Cut all the way through the meat until you hit bone. This will probably take more than one slice.

After you cut through all the meat on one side of the head, you need to switch to your bonesaw. You can use a large- or medium-sized bonesaw for this task.

3. **Holding your bonesaw in one hand and steadying the head with the other, use a soft arm motion to saw through the spine; stop sawing as soon as you saw through the bone (Figure 9-1c).**

When you use a *soft arm* (or *light sawing*) motion, you don't push the saw with all of your weight behind it because doing so can cause the teeth of the saw to anchor to the bone. Instead, use light force until a groove has been made in the bone and then put your weight behind it. The thickness of the bone determines how much force you put behind your bonesaw; you don't want to overshoot the bone and saw into muscle.

Now you just have to cut the head completely free of the body.

4. **With a larger butcher knife, slice through the remaining meat and cut the head free (Figure 9-1d).**

Don't throw the head away. You can use it to make head cheese, roast it for tacos, or cure the cheeks for *guanciale,* a cured, smoked pork jowl known for its sweet, smoky, complex flavor.

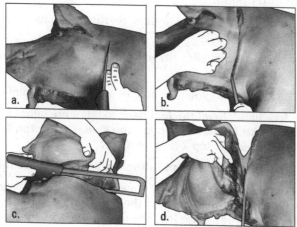

Figure 9-1:
Removing
the head.

Illustration by Wiley, Composition Services Graphics

Removing the Front Trotters (Feet)

You can remove trotters in one of two ways: Find the joint and use a boning knife to cut through it, or saw the foot off, using a combination of your boning knife and your bonesaw. Either way is correct.

Removing the trotters with your boning knife

Removing the foot or trotter with your boning knife helps you gain experience cutting through joints, but if you're a beginner, it may be a little frustrating. Don't feel discouraged if you hit some pesky cartilage that doesn't allow you to cut through the foot. Always be aware of your free hand, but use motion and observation of the joint as your solution to finding the right spot.

Follow these steps, shown in Figure 9-2:

1. **Bend the foot with one hand and feel inside the trotter where the joint is. Identify the opening where the joint connects and moves (Figure 9-2a); this is your entry point.**

   To give yourself some range of motion, let the legs hang off the table slightly.

2. **With your boning knife, slice through the thin flesh to expose the joint (Figure 9-2b and c).**

   If you hit the right mark, you'll be able to see where the joint connects.

3. **With your free hand, bend the foot back to open the joint even further. Then take your knife and cut around the side of the joint, following the line you started in Step 2 (Figure 9-2d).**

   This cut opens up the ankle joint further and allows you to slice through.

4. **Still holding the end of the foot in your free hand (be careful of that hand!), bend the foot back even further to open up the joint as much as possible. Insert your knife into the open joint and cut through with medium force to remove the foot (Figure 9-2e).**

5. **Repeat Steps 1-4 on the other foot.**

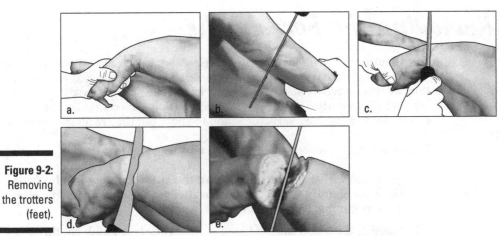

Figure 9-2:
Removing
the trotters
(feet).

Illustration by Wiley, Composition Services Graphics

Removing the trotters by sawing

Removing the foot or shank with a bonesaw is easier than removing it with a boning knife, and it also looks better. The cut is clean with no exposed joints, which can be a little unappetizing in a pot of beans or in the meat case.

To remove the trotters with a bonesaw, follow these steps (see Figure 9-3):

1. **Identify the ankle joint and slice through the skin around all sides of the ankle joint (refer to Steps 1 and 2 in the preceding section).**

2. **Using your bonesaw, saw through the bone (Figure 9-3a).**

3. **With your boning knife, cut through any remaining skin on the underside of the foot to cut the foot free (Figure 9-3b).**

Figure 9-3:
Removing
the trotters
with a
bonesaw.

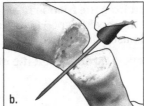

Illustration by Wiley, Composition Services Graphics

Removing the Foreshanks

The shanks are essentially the forearm (foreshank) or calf (hindshank) of the pig. These cuts are typically smoked (but also fabulous stewed) and are additionally recognized as ham hocks or pork knuckles. When smoked, they are known as *hocks;* when sold fresh, they're typically called *shanks.* The nomenclature for this cut may vary from shop to shop, but they're all the same cut.

To remove the foreshank, follow these steps (see Figure 9-4):

1. **Look and feel around to find where the elbow bone protrudes from the arm. Identify a straight line parallel to the spine above the elbow as your cut line (Figure 9-4a).**

2. **Holding the lower part of the arm with your free hand, cut through the meat directly above the elbow, following your cut line. Begin by cutting a straight line on top of the shoulder, above the elbow; then continue cutting through the flesh by moving your knife around the front and back end of the arm (where the elbow sticks out) (Figure 9-4b, c, and d).**

   Make sure to cut all the way through the meat until you hit bone.

3. **After you have cut through all the meat, saw through the arm bone (Figure 9-4e).**

   You can use gravity and your free hand to assist you by letting the arm "fall off" as you are sawing, enabling you to see where the bone ends and the flesh begins again. As soon as you're through the bone, stop.

4. **Use your boning knife to cut down through the flesh in a straight line and remove the shank completely (Figure 9-4f).**

5. **Use a kitchen towel to wipe the meat in the areas around both of the arms where you sawed through the bone.**

   When you saw through any bone, remnants of bone fragments and emulsified blood from the friction of sawing are left on the meat. This residue can cause the meat to spoil faster, so taking a moment to wipe the bone and blood paste off is worth it.

6. **Repeat Steps 1 through 5 on the other arm.**

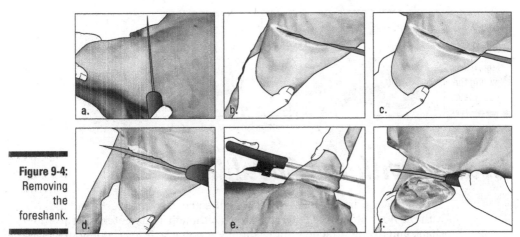

Figure 9-4:
Removing
the
foreshank.

Illustration by Wiley, Composition Services Graphics

Splitting the Breast-plate

Depending on the slaughterhouse's style (different processing facilities may process carcasses differently), you may need to split the breast-plate, or sternum. If the breast-plate of your animal is already split, you can skip this section and head straight to the next section, "Dealing with the Shoulder."

To split the breast-plate, follow these steps, shown in Figure 9-5:

1. **Starting at the belly and ending at the neck, use either your boning knife or a butcher knife to cut through the meat all the way down until you hit bone (Figure 9-5a).**

   The stomach has already been removed at the slaughterhouse, so you can continue this line up the center of the breast-plate. After you cut through the skin and meat, you're ready to saw through the breast-plate.

2. **Place your free hand on the top of the shoulder to steady the carcass, and using your bonesaw, start at the belly and saw down the center of the breast-plate inside the cut line you completed in Step 1 until you've sawed completely through (Figure 9-5b).**

   After you get about half way up the breast-plate, you may find standing near the belly side of the animal easier than standing near the front (the neck). This position lets you change the direction of your sawing, giving you the muscular advantage of pushing the saw in front of your body instead of pulling it up towards your body, and finish up the task by sawing through the section of the breast-plate closest to the neck.

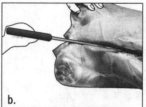

Figure 9-5:
Splitting the
breast-plate.

a.   b.

Illustration by Wiley, Composition Services Graphics

Dealing with the Shoulders

The shoulder primal is located between the neck and the fourth and fifth rib with the hock/shank typically removed. (Although some butchers leave the shank and trotter attached to the shoulder, doing so is less common because each piece lends itself to different cooking preparations.) To remove the shoulder, you cut a straight line between ribs 4 and 5 from the top of the shoulder down to the bottom edge of the rib cage. After that, you split the shoulders in two and do some final trim work, as the following sections explain.

Removing the shoulders

You remove the shoulder from the loin between ribs 4 and 5 for a reason: This is where the muscles change from being part of the shoulder to being part of the loin. As you move down into the loin, the muscles become leaner and more tender and are better suited for quick grilling and roasting than braising.

When you remove the shoulder, the tip of the shoulder blade is left on the loin, and the scapula (shoulder blade) and arm bones are left in the shoulder.

To make all the necessary cuts to the shoulder, follow these steps (see Figure 9-6):

1. **Looking at the rib cage from the inside of the carcass, count down the rib cage, starting at the first rib (the smallest rib near the neck), to ribs 4 and 5 (Figure 9-6a).**

   Look and look again. Sometimes rib 1 is covered by meat, fat, and dried blood, and can be hard to see. Using the point of your knife is a good way of checking. Tap until you hear the sound of bone and then count.

2. **Hold the place between the two ribs with your fingers on the inside and take a look at the outside of the shoulder as a visual reference.**

3. **Poke your knife through the rib cage and out through the other side of the shoulder (Figure 9-6b).**

You use this mark to identify the outer cut line you'll follow to remove the shoulder.

4. **Score down the rib cage on the inside of the ribs (Figure 9-6c).**

   Now that you know where to start, you can make quick work of cutting through the meat, because you simply follow the ribs as you saw through the spine and rib cage.

5. **From the top of the shoulder and holding your knife in a butcher's grip, cut down the shoulder in a straight line from the spine to the rib cage, following your cut line between the ribs (Figure 9-6d).**

   You may decide to switch to a butcher knife to cut through the thicker muscle on the top of the shoulder. Cut all the way through the skin and meat until you hit bone (the spine), which you have to saw through.

6. **Using your bonesaw, saw through the spine (Figure 9-6e).**

   Be careful not to saw past the bone into the shoulder meat on the other side of the hog.

7. **With a large butcher knife, cut through the shoulder and loin meat on the other side of the hog, continue to slice through the shoulder and down between the ribs 4 and 5 on the other side of the hog, and then cut through the sternum and down through the skin to remove the shoulders completely (Figure 9-6f through h).**

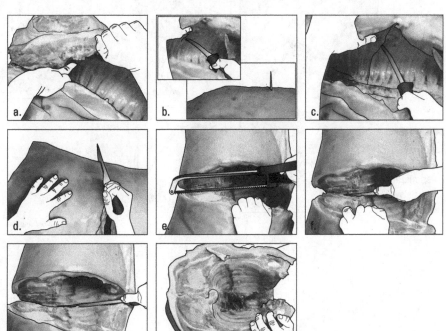

Figure 9-6:
Removing
the shoulder.

Illustration by Wiley, Composition Services Graphics

Splitting the shoulders in two

After you remove the shoulders, you're ready to split them into two individual shoulders. To do that, you saw down the center of the spine. Follow these steps (see Figure 9-7):

1. **Anchor the blade of your saw in the center of the spine; then pull the blade toward you slowly in a downward motion a few times to create a groove in the bone (Figure 9-7a).**

2. **Begin sawing, keeping your saw in the groove as you work (Figure 9-7b).**

   This task requires a little elbow grease, so when you find a comfortable groove, put your back into it.

3. **Saw completely through the spine, using your hands to pull down on the rib cage and helping you split the spine in half as you saw (Figure 9-7c).**

4. **Using a butcher knife or cimeter, cut through the meat in the center of the halved spine (Figure 9-7d and e).**

   Try to make a straight, clean cut, using a fluid motion (no ragged edges) and letting your shoulder to do the work instead of your wrist.

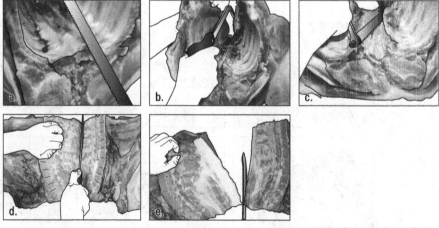

Figure 9-7:
Splitting the shoulders in two.

Illustration by Wiley, Composition Services Graphics

Trim work: Cleaning up the shoulder

After you successfully produce two shoulders from the pig carcass (hurrah!), you're ready to do a little trim work. As a butcher, you're responsible for trimming extraneous fat, glands, blood, or sinew from any cut you work on.

To clean up the shoulder, simply use the tip of your boning knife to trim the small bits of gland, blood, and connective tissue from the shoulder, up near the neck. You can use your free hand to pull up on the sinew while you carefully slice it away.

Although a little blood on the meat is not going to hurt anyone, blood that's settled on top of the meat can also make it taste bitter. Plus a cut is more likely to sell if it looks clean and well fabricated.

The hog may come to you with the offal (internal organs) still attached. Look inside the cavity and cut the kidneys free along with any sinew or connective tissue, using the tip of your boning knife. Set the offal aside for pâté.

Removing the Hind Trotters

At this stage, you've removed the head, two front trotters, two shanks, and two shoulders. Now you're left with the loin, belly, and legs. Time to remove the two hind trotters. Follow these steps (see Figure 9-8):

1. **Locate the front of the knee joint; then using your boning knife, score around the bone and cut through the skin on the top of the leg (Figure 9-8a).**

   Taking the hind trotters off is similar to removing the front trotters, but the joint is shaped differently. Keep this in mind as you examine the joint to locate your cutting point.

2. **With your bonesaw, saw through the trotter, using the cut you made in Step 1 to identify your saw line (Figure 9-8b); stop when you're through the bone.**

3. **Use your knife to cut the trotter completely free (Figure 9-8c).**

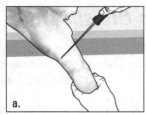  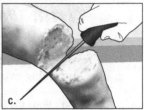

Figure 9-8:
Removing
the hind
trotters.

a.   b.   c.

Illustration by Wiley, Composition Services Graphics

Sectioning the Legs from the Loin

Now that you have removed the hind trotters, you are left with two legs (shanks attached) and the saddle (or loin) with the belly still attached. So it's time to remove the legs from the loin in a two-step process: First you cut the belly free, and then you remove the legs from the loin, as the next sections explain.

Freeing the legs from the belly

Follow these steps to separate the belly from the legs (see Figure 9-9):

1. **You may need to cut through skin and meat at the bottom of the pelvis. If so, follow the existing cut in the center of the belly to cut through the skin and meat until you reach the tail (Figure 9-9a).**

   If the lower half of the belly is already split to the tail, skip to Step 2.

2. **Open the stomach cavity and look inside the loin to find the two tenderloins running on either side of the spine (Figure 9-9b).**

3. **Run your fingers along the spine close to where it curves into the tail to find the vertebrae; insert your knife into the disc between the vertebra just above the tail (Figure 9-9c).**

   Look closely to see each vertebra. You can identify a vertebra by two raised bumps close to each other with a white space (the disc) between them.

4. **Rock your knife forward and backward to cut further down into the vertebra and through the tenderloin (Figure 9-9d).**

   This prep cut ensures easy removal of the loin after you have completed the next few steps.

Following these instructions, you *are* cutting through the tenderloin. If your end goal is to cut porterhouse or T-bone steaks from the loin, this is the correct way. If you want to remove the tenderloin in one piece, you may do so before performing this step. Chapter 10 explains how.

Now you're ready to cut the belly free from the leg.

5. **With your free hand, grab the lower part of the flank or belly (near the leg) and cut down the flank, slicing to separate it from the leg, angling down the front of the leg until you reach the belly (Figure 9-9e through g). Stop cutting when you connect with the cut line you created between the vertebra in the loin.**

6. **Repeat Step 5 on the other side of the carcass to release the belly from the legs on both sides.**

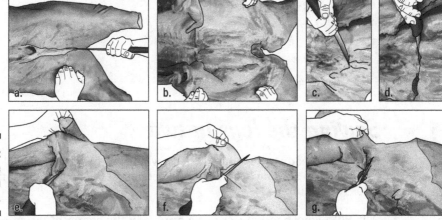

Figure 9-9:
Freeing the legs from the belly.

Illustration by Wiley, Composition Services Graphics

Separating the loin from the legs

Now you're ready to separate the loin from the legs (see Figure 9-10).

1. **Turn the carcass on its side. With a large butcher knife or cimeter, make a straight cut on the side of the loin and around the back, following the same line you created between the vertebra; repeat this cut on the other side (Figure 9-10a and b).**

At this point, all the meat on either side of the leg and parallel to the separated vertebra should be cut and ready for the bonesaw.

2. **Saw through the spine, inside the cut line (Figure 9-10c). When you're through the bone, stop.**

3. **With a large butcher knife or cimeter, cut through the remaining meat on the underside of the loin until the loin is free (Figure 9-10d and e).**

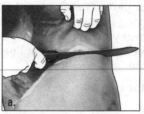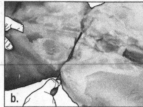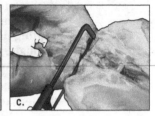

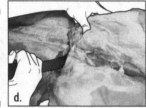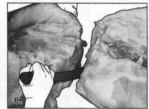

Figure 9-10: Separating the legs from the loin.

Illustration by Wiley, Composition Services Graphics

Sawing the legs in two

Now you're ready to separate the legs, which, at this point, are still connected. To do so, simply saw down the center of the spine. Follow these steps (see Figure 9-11):

1. **With your bonesaw, create a groove in the center of the spine (Figure 9-11a).**

2. **Continue sawing until you have passed through all the bone; then stop (Figure 9-11b).**

   You can use your free hand to push the legs open, helping to split the spine apart as you work.

    Don't bother trying to saw the tail in half; it's better left in one piece. Instead, saw on one side or the other.

3. **Using your large butcher knife or cimeter, cut the legs in half down the center in a strong, straight line (Figure 9-11c and d).**

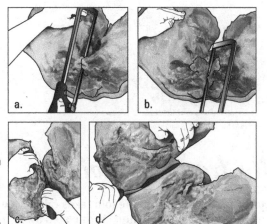

Figure 9-11: Sawing the legs in two.

Illustration by Wiley, Composition Services Graphics

Removing the Pork Skirt Steaks

Before you make your finishing cuts, you need to do a little trim work inside the saddle. Here, you find two pork skirt steaks (the diaphragm) attached to the inside edge of the rib cage on either side. These steaks are very easy to remove. You can set them aside for grinding, or if you have a little patience, you can clean them.

After you remove the skirt steaks, spend a moment cleaning any extraneous blood, fat, or sinew from the inside of the rib cage as well. Doing this task now saves you time later.

To slice the skirt free, follow these steps (see Figure 9-12):

1. **Find the skirt steak located on the inside of the belly.**

   The skirt steak looks like a flap of meat on either side of the saddle and is located where the rib cage ends, away from the spine.

2. **With your free hand, grab the top edge of the skirt steak; then use your boning knife to make a slice at the top where the skirt is attached to the rib cage (Figure 9-12a).**

3. **Slice down from the top to the bottom of the skirt, removing the skirt steak in one whole piece (Figure 9-12b).**

4. **Repeat these steps to remove the skirt steak on the other side.**

Figure 9-12:
Removing
the skirt
steaks.

Illustration by Wiley, Composition Services Graphics

Cutting the Belly from the Loin

Time to finish up! Removing the belly from the loin takes only a few final cuts and a little sawing. At this stage, the animal is one step away from being broken into organized muscle groups for easy storage and portioned for future use. By following the steps in the next sections, you'll have two bellies and one double loin that you can split later.

Follow these steps to be well on your way to smoking bacon and grilling up pork chops (see Figure 9-13):

1. **Take a look at the spine to get your bearings. Estimate about 3 inches of rib bone and take note of where the strip loin ends on the opposite side.**

   On either side of the spine are the ribs that are part of your bone-in pork chops. If removed, they become baby back ribs.

   Your task is to divide the belly from the loin, leaving about three inches of uniform ribs on either side of the spine. You also need to pay attention to the meat on the opposite end of the loin — the *strip loin* (think New York strip on a cow) — as you determine your cut line. 'Twould be a shame to slice into it.

2. **Starting with the strip loin, measure 1 inch past where the strip loin muscle ends and make a small notch (Figure 9-13a).**

3. **Turn the saddle so that the rib end is closest to your body. Then using the notch you made on the other side, draw your knife toward you, cutting through the belly meat and across the rib bones (Figure 9-13b).**

   Make sure your cut line is straight and uniform. This cut prepares the belly for sawing. You want to know that the belly is portioned properly before you start to saw; this way you have a predetermined guideline for sawing through the rib bones. You are essentially drawing a line to cut on.

4. **Use your knife to score down the ribs as you finish this cut (Figure 9-13c).**

   You'll use this guideline when you saw through the ribs. If you can't see the line very well, run your knife over the ribs a couple more times.

5. **Saw through the ribs following your cut line. When you get through the rib bones, stop (Figure 9-13d); don't cut through the meat below.**

   The ribs are fairly thin, about 1/4-inch thick (depending on how large the hog is, possibly thicker).

6. **Holding your butcher knife or boning knife in a butcher's grip, cut the belly free from the loin through the center of the split ribs, connecting with your cut line near the strip loin, to remove the belly in one square piece (Figure 9-13e).**

   Your cut should be strong and straight. No wobbly wrists or hacking to remove the meat. If you don't cut all the way through the skin and meat during this first pass, don't let it stop you. Finish the slice, then go back over with the same motion, following the same line and repeat the cut.

   Last, but most definitely not least, you need to make sure your belly is looking super professional before you switch to the other side. Take a moment to square up the belly and feel proud of your work.

7. **Using a large butcher knife or cimeter, square up the edges of the belly (Figure 9-13f) and set the trim aside for grinding.**

8. **Repeat Steps 1 through 7 to cut the belly from the loin on the other side.**

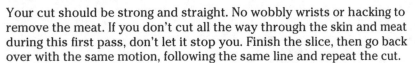
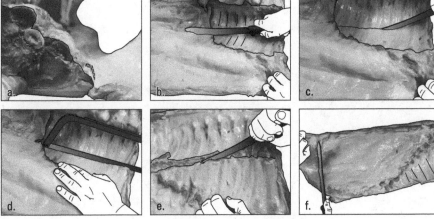

Figure 9-13: Cutting the belly from the loin.

Illustration by Wiley, Composition Services Graphics

Chapter 10

Moving into Pork Subprimals

. . . .

In This Chapter

▶ Fabricating retail cuts from the butt and shoulder

▶ Producing cuts from the loin and leg

▶ Loading up with helpful advice on how best to trim pork

. . . .

After you have butchered a carcass into primals, as I explain in Chapter 9, the next stage is to cut it into subprimals. Subprimals are conveniently small enough to transport in whole pieces and yet large enough to preserve the whole muscle until you're ready to portion the meat into retail cuts to be cooked or placed in the meat case and sold. For this reason, subprimals are the standard wholesale method of sale to restaurants and butcher shops.

After you have broken the primals into subprimals, you can then create the retail cuts, which are simply cuts of meat sold in a retail market, generally portioned for one meal. Here's where the fun and confusion come in: Depending on how it's cut, a subprimal can yield different types of retail cuts. Popular pork powerhouses like the belly give you bacon, or pancetta, spareribs, and a bevy of braised bounty. The loin houses the popular center-cut pork chop, baby back ribs, boneless pork loin, T-bone steaks, or tenderloin. Unfortunately within the loin, you can't have it all: This is the stage where choices must be made. Do you want center-cut bone-in pork chops or baby back ribs? You decide. The instructions in this chapter produce most standard retail cuts.

To make the cuts outlined in this chapter, you must have completed the butchering outlined in Chapter 12. The cuts I include here are based on the primal cuts made there.

From the Shoulder: The Boston Butt and Pork Shoulder (Picnic)

Within the shoulder primal you have two subprimals: the Boston butt and the pork shoulder, or picnic. The Boston butt is well marbled, has great flavor, and, when braised, is a delicious, rich, and reasonably priced meat. The picnic has less fat and more connective tissue, and is known to be a tougher cut (the picnic is the lower portion of the shoulder, and its muscles work harder than the Boston butt muscles, which are on the upper shoulder — makes sense, right?). Despite being a less tender piece of meat, the picnic is delicious braised, stewed, or prepared like a ham.

Some resources list the Boston butt and the picnic as primals themselves; others identify them as subprimals of the shoulder, as I do here. Conflicting information about butchery techniques and cuts is common, so don't let it confuse you. Just remember that the natural progression of the butchery process is to go from large pieces (primals) to smaller pieces (subprimals) to individual cuts (retail cuts).

Separating the Boston butt from the picnic

Separating the Boston butt from the picnic is fairly simple. You need a boning knife, a butcher knife, and a bonesaw. From the Boston butt, you can produce Boston butt steaks and bone-in and boneless roasts.

Many of the thin bone-in steaks outside of the loin are difficult to cut if you don't have a bandsaw, so if you're butchering by hand with a bonesaw, I don't recommend trying to cut bone-in butt steaks. Choose the boneless version instead.

Partially deboning the shoulder

To separate the butt from the picnic, you need to debone the remaining ribs, spine, and neck bones from the shoulder. This task is the trickiest part of separating the butt from the picnic, but, if you followed the instructions in Chapter 12, you already know the basic technique.

Follow these steps, shown in Figure 10-1:

1. **Lay the shoulder on the butcher block, skin side down with the neck facing away from you. Then trace around the breastbones with your boning knife (Figure 10-1a).**

   The breastbones are the bones opposite from the spine, on the belly side of the pig.

2. **After you trace around the bones, pull up on the breastbones a bit with your fingers and slice down against the ribs (Figure 10-1b).**

   Doing so allows you to make some space between the meat and bone. Use this space as your access point, while slicing down against the ribs.

3. **Continue to slice down, against the ribs, while holding the breast bone with your free hand (Figure 10-1c).**

   As you cut down either side of the ribs and get closer to the spine, you'll find that the section of ribs and breastbone are free from the meat. Grab hold of the section of ribs and use this as a handle while you continue your deboning work around the spine and neck bones.

   At this stage, it may help you to turn the shoulder about 75 degrees and position yourself behind the bones. With the belly side facing away from you, pull back on the bones with your free hand and use the tip of your boning knife to continue working around the spine.

4. **Cut the rib, spine, and neck bones free from the shoulder meat (Figure 10-1d and e).**

   You use the techniques explained in the preceding chapter to work your way around the rib and collar bones. Keep your knife flush against the bone (and always pay attention to your where your fingers are) while working to free the bones from the flesh. Basically you want to cut as closely along the outline of the bones as possible to get as much meat as you can.

   If you get frustrated or feel that you're leaving too much meat, stop. Put your knife down, feel around the bones with your fingers to get a feel for the muscles, and then start again. And remember: If you end up with a little too much meat left on the bones, don't be too hard on yourself. The way I see it, you have a magical pot of beans or a solid stock in the making. Don't forget to set the bones aside for later.

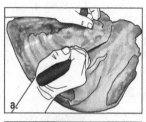

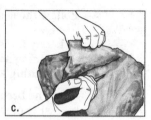

Figure 10-1: Removing the rib, spine, and neck bones from the shoulder.

Illustration by Wiley, Composition Services Graphics

Separating the butt from the picnic

After you remove the spine, rib, and neck bones from the shoulder, the hard part is over. The rest is just a matter of determining where to cut and then making one solid cut and doing a little saw work. Follow these steps, shown in Figure 10-2:

1. **Set the shoulder on the block, skin side up and with the shoulder turned so that the spine is on the left side of your body, the arm on the right, and the blade end — the end where the shoulder blade was — facing away from you (Figure 10-2a).**

   Butchery resources use terminology like *blade end, rib end, sirloin end,* and so on quite often. Consider these the meat equivalent to terms like *fore* and *aft* because they help you locate which side of the subprimal is being addressed. The blade end is where the shoulder blade is/was located.

2. **Size up your cut line down the center of the shoulder.**

   You're trying to cut the shoulder in half, so make sure the size of the shoulder on either side of your cut line is more or less equal.

3. **Using a butcher knife or a boning knife, make a straight cut, parallel to the spine, down the center of the shoulder (Figure 10-2b).**

   Focus on making a straight, clean cut, using a fluid, strong arm while cutting all the way through the meat.

4. Saw through the bone (Figure 10-2c).

5. After you have sawed all the way through the shoulder blade, switch back to a butcher knife and cut through the remaining meat to separate the butt from the picnic (Figure 10-2d).

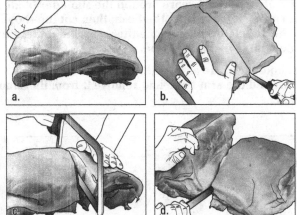

Figure 10-2: Separating the butt from the picnic.

Illustration by Wiley, Composition Services Graphics

Deboning the Boston butt

Time to get a little deboning practice in on the pork butt. You can use your new skills to debone the picnic, described in the next section.

When you're new to deboning, unexposed bones inside larger sections of muscle can seem intimidating. So always take a moment to look at the bones on either side of the subprimal. Notice where they both end and see whether you can discern where they connect to the each other or to other bones. Visualizing the location and direction of the bone inside the shoulder helps you feel more confident when you go to remove it.

To debone the pork butt, follow these steps, shown in Figure 10-3:

1. **Make a 3-inch cut in the center of the butt, from the top of the blade end, directly on top of the shoulder blade (Figure 10-3a).**

   Your knife should be hitting bone as you make this cut.

2. **With your boning knife, make a cut that follows the bone from the shoulder blade down to the end of the arm bone (Figure 10-3b).**

   As you make this cut, be sure to keep your knife against the bone.

Now that you have sliced the shoulder open a little, you can see more clearly where the bone is inside the shoulder.

3. **Keeping your knife against the bone, make small slices as you pull the meat away from the bone with your free hand.**

    This cut opens up the shoulder and exposes the "top-side" of the bone.

4. **With the tip of your knife, score around the shoulder blade and the arm bone, cutting it loose from the flesh; then cut the shoulder blade free from the muscle by slicing on either side and under the shoulder blade while pulling up on the bone.**

5. **Using the shoulder blade as a handle and grabbing it with your free hand, cut around the arm bone and remove it from the shoulder (Figure 10-3c).**

Figure 10-3:
Producing
a boneless
pork butt.

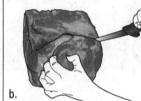
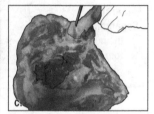

a.   b.   c.

Illustration by Wiley, Composition Services Graphics

Making retails cuts from the picnic

At this point, you have separated the picnic from the butt. You've already removed the shank from the picnic above the knee joint (refer to Chapter 12), and you've separated the pork butt with a cut running parallel to the spine.

From the picnic, you can produce a variety of boneless and bone-in retail cuts: an outside shoulder, boneless inside shoulder, pork cushion meat, pork brisket, smoked hocks, arm picnic roast, or a boneless picnic shoulder. All these cuts can be tied for roasts, cured and smoked, cubed into stew meat, or ground to be used in sausage-making and further processing. In my opinion, it's best to debone the picnic (refer to the preceding section on deboning the pork butt) and cube it for stew meat, grind it, or tie it for some really nice, smaller roasts (which are best brined).

Boneless picnic shoulder cutlets

After you debone the picnic, you can cut thin pork cutlets. Keep in mind this muscle is generally tougher, so you may want to brine, tenderize, bread, or marinate these shoulder steaks before cooking.

To produce picnic shoulder cutlets, you simply trim the picnic shoulder free of any sinew, lymph glands, or seam fat, and then cut it into steaks that are no thicker than a ¼ inch.

When you shop for meat, you may be confused by the distinction between a cutlet and a chop. Here's a quick primer: The term *pork cutlets* applies to *any boneless,* thinly sliced pork steak that doesn't come from the loin. Cutlets are typically fabricated from the leg or shoulder. All the pork steaks that come from the loin are called *chops:* loin chops, ribs chops, boneless pork chops, and so on.

Skinless pork shoulder hocks

Skinless pork shoulder hocks are great for braising or stewing, if you don't plan to smoke them. Smoked hocks are fabulous cooked in stews or in a pot of beans to add condensed flavor.

If you followed the instructions in Chapter 9, in which I explain how to separate the shoulder hocks (shanks), you need only remove the skin. Follow these instructions, shown in Figure 10-4:

1. **Make a straight slice through the skin from top to bottom (Figure 10-4a).**

   Try to slice only through the skin without cutting into the meat.

   The skin can be tough to cut through, so try not to struggle with it. That's how knives slip and fingers get cut!

2. **With your free hand, pull up on one corner of the skin; then use the tip of your boning knife to slice under the skin as you pull the skin back (Figure 10-4b).**

3. **Continue cutting in this manner, slicing under the skin and pulling the skin away from the hock, and turn the hock as you cut (Figure 10-4c).**

4. **Cut the skin free from the hock (Figure 10-4d).**

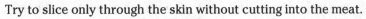

Save the skins to use for chicharrones (see Chapter 16 for a recipe), dry them for dog chews, or store them in a resealable freezer bag for later use.

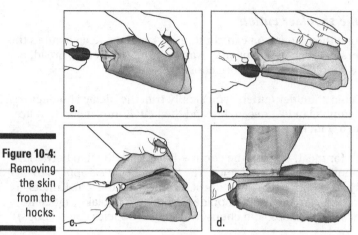

Illustration by Wiley, Composition Services Graphics

Figure 10-4:
Removing
the skin
from the
hocks.

Producing Retails Cuts from the Loin

The loin houses many of the "center of the plate" cuts you're probably most familiar with: boneless pork loins, loin roasts, bone-in pork chops, baby back ribs, pork tenderloin, pork sirloin steaks, sirloin roasts, boneless pork strip loins, pork porterhouses or T-bone steaks . . . the list goes on. These cuts are well-suited for quick cooking methods like grilling, pan searing, or roasting, and they're usually sold as individual portions. The muscles in the loin do less work than the muscles in the shoulder or the leg do, so they're more tender. Some cuts are well marbled, whereas others are lean. Without a doubt, the loin offers versatility in terms of variety of cuts and preparation possibilities.

Cutting center loin chops

Center-cut loin chops are located between the sirloin and the rib end. They're bone-in chops with tenderloin and top loin cuts on either side of this classic T-bone cut. The instructions in this section follow the progression of butchery from the previous chapter, where you cut the belly from the loin.

Removing the chine (whole hog)

Because you started with a whole hog rather than a side of pork, you need to split the rack into two first (that is, you need to remove the spine so that you have two loins). *Note:* If you are starting with a side of pork, you can skip this section.

The *chine* (spine) is a pesky piece of bone when you're ready to cut chops. Although you can separate the chops by cutting in between each vertebra or sawing through the chine at the point you intend to portion the chops, I recommend that you remove the chine completely because doing so allows you to easily cut chops.

To remove the chine, follow these steps, shown in Figure 10-5:

1. **Lay the center loin skin side down with the spine perpendicular on the cutting board in front of you. Then on either side of the spine, identify where the spine ends and the ribs attach to the spine.**

2. **With your knife, score a parallel line along each side of the spine (Figure 10-5a).**

   Run your knife down each side several times to identify your cut line and create a groove for your bonesaw during the next step.

3. **Place the bonesaw inside the groove on one side of the chine and saw in between the rib bones and the spine at a 45-degree (inward) angle (Figure 10-5b).**

   Be very careful not to saw past the bone and into the muscle. Keep your eye on the visible end of the spine so that you can gauge how thick the bone is and stop sawing before you reach meat.

4. **Turn the loin 180 degrees and repeat Steps 2 and 3 on the other side (Figure 10-5c).**

5. **With a boning knife, cut down either side of the chine, making sure that your saw work releases the spine from the ribs (Figure 10-5d). With your free hand, lift up on the spine, pulling it up and away from the racks, and cut it free from one side of the chine.**

   If you failed to saw all the way through the spine the first time, pick up your saw and finish the task.

6. **In a straight line in the center point between the rack and parallel to the spine, cut through the skin and separate the racks into two individual pieces (Figure 10-5e).**

7. **Cut the chine free from the other rack.**

8. **With a clean towel, wipe the meat clean of any blood or bone dust.**

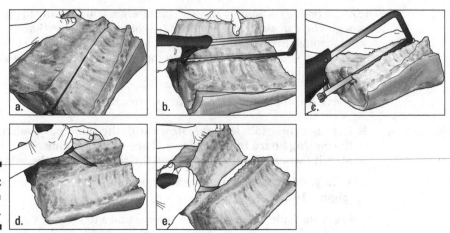

Illustration by Wiley, Composition Services Graphics

Figure 10-5:
Splitting the
rack.

Removing the skin

Time to remove the skin. Follow these steps, shown in Figure 10-6:

1. **Lay the rack rib side down on the cutting board in front of you.**

2. **Slice under the skin, with a slow, solid, smooth motion from the top edge to the bottom (Figure 10-6a).**

   Try not to saw. A sawing motion creates jagged marks in the fat under the skin.

3. **Lift up the skin, identify where the skin is still connected, and continue cutting under the skin until the skin is removed completely (Figure 10-6b).**

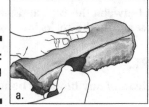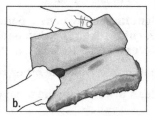

Figure 10-6:
Removing
the skin.

Illustration by Wiley, Composition Services Graphics

Portioning the center loin chops

Now that all the prep work is done, cutting the center loin chops is an easy task. You can cut the chops any desired thickness: single, bone-in chops; double rib chops; or even thin chops alternating between bone-in and boneless.

Regardless of the size chop you're cutting, remember to use a large knife and make deliberate cuts, dragging your knife through the meat in one solid motion (see Figure 10-7). Doing so ensures that your chops are straight and attractive. (Be careful not to move the loin with your free hand, or you may end up with weird shape and uneven thickness.)

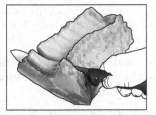

Figure 10-7: Portioning your loin chops.

Illustration by Wiley, Composition Services Graphics

Going for country-style ribs

Country-style ribs are produced from the blade end of the loin. They're great for grilling or pan roasting. This standardized retail cut (the chine is removed prior to fabrication) has between three and six ribs. The section of ribs is then butterflied down the center of the loin. After butterflying, the ribs are cross-cut between each rib to produce several individual, meaty portions.

If you intend to cut country-style ribs, set aside the three- to six-rib blade end section while you portion the loin (before you cut rib chops). Or if you've already cut the whole loin into chops, you can butterfly them individually.

Boneless loin roast and chops

Many chefs and butchers would agree that removing the bones from a center-cut pork loin is a waste. The rib bones have a lot of great meat and fat on them, and the bones add flavor during cooking. But if you are looking for a lean roast or would prefer to have baby back ribs over bone-in rib chops, deboning the blade end of the loin is the way to go. The instructions in this section explain how to produce a boneless center-cut loin roast. You can also produce boneless loin chops by cutting them into individual portions.

Deboning the rib end and center loin

Both the center loin and the strip loin (or rib end loin) can be processed as boneless roasts. In fact, the whole loin can be deboned, a very common way to process the loin. But for the purposes of this section, I explain how to debone the rib end and center loin. Follow these steps, shown in Figure 10-8:

1. **Place the bone-in rack chine down on the cutting board in front of you.**

2. **Slice against the rib bones, starting at the top. Pull your knife across the bones from front to back, making deliberate slices, keeping the pressure of your knife against the bones. Cut down the ribs about an inch or so (Figure 10-8a).**

Leaving some meat on the rib bones while you are slicing down the ribs is okay. Normally when you debone, the goal is to cut the bones clean (free of meat), but in this instance, you'll be cutting baby back ribs from the leftover bones. Meaty ribs are good.

3. **Pull back on the cap with your free hand, simultaneously slicing against the bone and pulling the meat away, as you cut down the rack. Cut until you reach the chine (Figure 10-8b and c).**

The *cap* is rib lifter meat or wedge meat located on top of the rib.

4. **When you reach the chine, score against the chine and featherbones and then cut the loin muscle free (Figure 10-8d).**

At this point, you can either cut the meat into chops or tie it as a boneless roast.

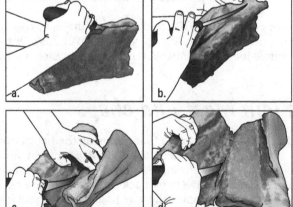

Figure 10-8:
Deboning the center loin.

Illustration by Wiley, Composition Services Graphics

Tying a boneless pork loin roast

Although the muscles in the loin are fairly even in shape and size, you still need to tie them. Tying roasts serves two functions: First, it holds the meat together in a uniform shape, guaranteeing even cooking and ease of slicing after cooking. Second, when done properly, it makes oddly shaped muscles look attractive. Plus nothing says "roast" better than butcher's twine.

To tie a roast, follow these steps:

1. **Lay the section of boneless loin on the cutting board in front of you.**

2. **Place the twine next to you on the table and grab the cut end of the twine in your right (or left for left handers) hand.**

3. **With your free hand, grab the twine 12 inches or so down from the end and hold it taught. Slide it under the boneless loin and stop when you reach the center.**

   The hand holding the cut side of the twine should be farthest from your body.

4. **Tie the twine and cinch up on the string, making sure the twine is pulled tight before you complete the knot.**

5. **Divide the untied portion of the roast in half and tie again.**

   By beginning in the center for the first tie and dividing the untied portion of the roast into halves for the subsequent ties you ensure that your strings are evenly spaced and looking good! Nothing is worse than getting to the end of a roast and realizing that your strings are all wonky — the horror!

6. **Continue adding strings, tying down the roast in this manner until you reach the ends (see Figure 10-9).**

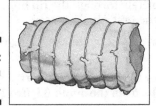

Figure 10-9:
Tying a
roast.

Illustration by Wiley, Composition Services Graphics

Boneless loin chops

If you'd rather have chops than a roast, you simply have to decide how thick you want your chops and then slice across the loin with a large butcher knife or cimeter. Make your cuts nice and smooth.

Baby back ribs

Champions of barbecue, smothered in sweet, tangy sauce and grilled, or pan seared and simmered in a spicy tomato salsa, baby back ribs are indisputably delicious and a favorite of cooks everywhere. Getting your hands (and face) dirty while gnawing on some tasty rib bones is just plain soul-satisfying.

By following these steps, shown in Figure 10-10, you produce one rack of baby back ribs (note that you need a boning knife, cleaver, and mallet):

1. **Lay the rib bones (with chine and feather bones still attached) bone side down on the cutting board.**

2. **At the base of the spine, use your boning knife to score the meat and bones (Figure 10-10a).**

   Essentially, you're "drawing" a straight line with you knife, parallel to the spine.

   Scoring the meat and bones in advance helps you identify where you'll cleave the bone, ensuring that you are produce a nice, rectangular slab of baby back ribs.

3. **Starting with the rib closest to your body, place the tip of your meat cleaver directly on top of the rib, inside the cut line you scored in the preceding step. Give the front of your meat cleaver a solid whack with your mallet to sever the first rib (Figure 10-10b).**

   If one whack does not make it completely through the rib bone, give it another, or until you have severed the first rib.

4. **Move down the ribs with your cleaver and mallet, severing each rib in turn, until you reach the end.**

5. **Use a large butcher knife or cimeter to cut between the ribs and the chine to separate the baby back ribs from the spine (Figure 10-10c and d).**

   Be sure to cut through any thin bits of bone or meat, as necessary.

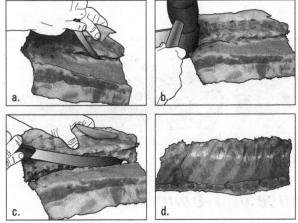

Figure 10-10:
Producing
baby back
ribs.

a.

b.

c.

d.

Illustration by Wiley, Composition Services Graphics

Removing the tenderloin

The tenderloin rests on either side of the spine at the ham end of loin. If you did not cut through the tenderloin when you freed the legs from the belly (refer to Chapter 9), you still have an intact, whole muscle that you can easily remove in one piece. In fact, you can do much of the work with your hands. Follow these steps, shown in Figure 10-11:

1. **Run your fingers along the spine, where the tenderloin is attached, pulling the muscle away from the chine (Figure 10-11a).**

2. **Score along the spine with your boning knife, continuing to pull the muscle away with your hands as you work (Figure 10-11b).**

   At this point, the tenderloin falls away from the spine but is still lightly attached to the loin by muscle and connective tissues (called the *chain*).

3. **Grab the tail of the tenderloin and lift up. Lightly slice under the tenderloin, pulling up as you cut (Figure 10-11c).**

4. **Work your way down the chain, removing the tenderloin from the loin in one whole muscle.**

5. **Carefully clean the tenderloin of fat and silver skin.**

You can cut the tenderloin into medallions or tie it (or stuff and tie it) to use as a whole roast.

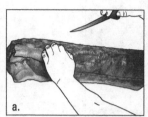 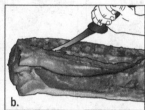 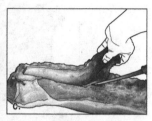

Figure 10-11:
Removing
the tender-
loin.

a. b. c.

Illustration by Wiley, Composition Services Graphics

Porterhouse or T-bone steaks

T-bone steaks are cut from the lower half of the loin, sirloin end (also known as the *short loin*). Great for grilling or pan searing, porterhouses house both the tenderloin and the loin on either side of the spine.

To create porterhouse or T-bone steaks, follow these steps, shown in Figure 10-12. (***Note:*** If you are starting with a side of pork, skip Steps 1 and 2.)

1. **Saw through the spine, using your bonesaw (Figure 10-12a).**

   Remember to anchor your bonesaw on one end of the spine by establishing a groove in the spine bone before going full-force with the saw.

2. **With a large butcher knife, cut down the center of the spine and through the skin to split the short loins in two (Figure 10-12b).**

3. **Remove the skin (Figure 10-12c).**

   For instructions, refer to the earlier section "Skinless pork shoulder hocks."

4. **With the bone-in loin chine down on the cutting board, use a large butcher knife or cimeter and make evenly portioned cuts through the meat, corresponding with each point on the spine where the vertebra connects (in between the discs). Cut down the entire length of the loin (Figure 10-12d).**

   Pay attention to the meat on the other side of the spine.

5. **Place your cleaver between the first and second chop and give it a good whack with the mallet to knock the chop right off (Figure 10-12e).**

6. **Continue cutting off the chops as you did in Step 5 until you are finished.**

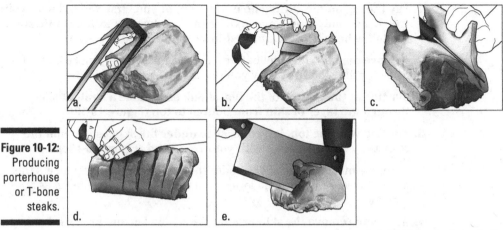

Figure 10-12: Producing porterhouse or T-bone steaks.

Illustration by Wiley, Composition Services Graphics

Getting Great Cuts from the Leg (or Ham)

The pork leg, or fresh (green) ham, is grossly underutilized and under-appreciated at the dinner table in the United States. Legs are primarily used for hams (or cured for prosciutto), but the culinary creativity usually stops there — an unfortunate situation. Although leg meat may be leaner and dryer than the loin, it's also inexpensive and responds beautifully to brining or curing and roasting. Because the leg has a lot of meat, you don't have to choose between a Sunday roasted ham and Tuesday night Milanese. You can have it all.

You can roast fresh hams whole or seam them out into smaller muscle groups to create several smaller fresh hams. In this section, I explain how to prepare a fresh ham.

A ham has enough skin to make a good-sized batch of chicharrones or pork cracklings. But it's also deliciously crispy when left on and roasted with the ham. Depending on how you prefer your ham, you may prepare the ham skin off or on.

Follow these steps to remove the tail and to remove the skin in one whole piece, shown in Figure 10-13:

1. **Lay the leg on the cutting board in front of you, grab the tail with your free hand, and, using your boning knife or butcher knife, cut through the tail, where it's still attached to the spine (Figure 10-13a).**

   Remove the tail with a deliberate slice, leaving it in one piece and set it aside for the stock pot. Now you're ready to remove the skin.

2. **Starting at the bottom of the leg, shank end, make a straight slice through the layer of skin from bottom to top (Figure 10-13b).**

3. **Starting with the top, left edge, slice under the skin as you pull the skin away from the meat with your free hand (Figure 10-13c).**

   Continue making small slices under the skin, where the skin meets the muscle as you pull the skin away from the meat, turning the leg as you go.

4. **After you remove the skin around the whole leg and are now back at the point where you started, cut the skin completely free from the leg (Figure 10-13d).**

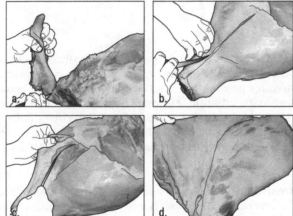

Figure 10-13: Preparing a fresh ham.

Illustration by Wiley, Composition Services Graphics

Of course, a pork leg can produce more than hams alone. You can also get pork cutlets, cube steaks, stir fry, kabobs, county hams, stew meat, ham center slices, shank roast, and leg roast.

Spareribs from the Pork Belly

Pork belly has become a hot commodity with chefs in recent years. The pork belly is most commonly known for the sweet and savory miracle that is bacon! Wonderful when cured and smoked (bacon) or cured and rolled (pancetta), pork belly is also equally enjoyable when slow roasted or braised.

When removed from the loin, the belly still has the spare ribs attached to it. Depending on what you want, you can remove only the ribs, leaving the intercostal meat on the belly. But if you're in the mood for barbecue, slice off a slab of spareribs and fire up the grill.

The spareribs are located on top of the belly. Follow these steps, shown in Figure 10-14, to remove and trim your spareribs:

1. **Lay the belly on the cutting board, skin side down. If the skirt steak is still attached to the end of the ribs, trim it off (Figure 10-14a).**

2. **Score around the inside edge of the ribs, cutting through the flesh no more than ½ inch (Figure 10-14b).**

   Make sure not to cut too deeply into the belly.

3. **With a boning knife or large butcher knife, slice under the ribs (Figure 10-14c).**

   Place your free hand on top of the ribs to steady the muscle as you cut.

4. **Pull the ribs up and slice under them until you reach the ends of the ribs (Figure 10-14d).**

   The scoring cuts you made in the Step 2 help you remove the spareribs easily in the next step.

5. **Pull up on the ribs and slice the spareribs free from the belly (Figure 10-14e).**

   Now you're ready to do some additional trim work.

6. **Identify a cut line parallel to the other side of the ribs. With your knife, score down the bone and cut through the cartilage (Figure 10-14f).**

   Make sure that the width between the ribs is even from top to bottom.

7. **Using the tip of your cleaver and your mallet, cut through the ribs one at a time, following the cut line you identified in Step 6 (Figure 10-14g).**

8. **Trim the uneven tail end off the ribs (Figure 10-14h).**

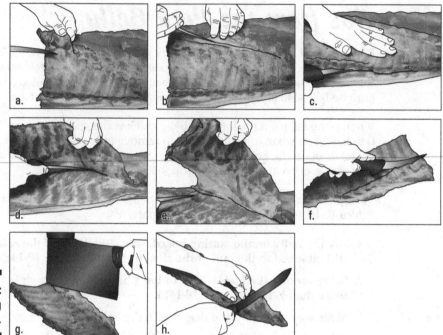

Figure 10-14:
Preparing
spareribs.

Illustration by Wiley, Composition Services Graphics

Trimming Meat for Grind

A meat grinder may not be standard issue in a typical home kitchen, but if you like butchering or making sausage, I highly recommend having one because a grinder opens up a whole world of culinary possibilities to you.

When you trim meat for grind, cut away any darkened meat (which has been *oxidized,* or exposed to air), glands, sinew, blood, or connective tissue. Don't worry about removing all the fat (you want fat!) or silver skin from the meat, but keep in mind that the sinew or connective tissue that's too thick or tough can wrap itself around the cutting blades of the grinder and put stress the grinder's motor. It can also lower the quality of your grind.

Matthew Jennings of Farmstead, Inc., in Rhode Island shared a great grinding tip with me: "Instead of cubing the meat for grind, as is generally taught, cut the meat into long strips about an inch in width. The opening of the grinder is long and about an inch wide. These strips feed through the grinder much more easily than cubes do, and they're faster to prep. Easy in, easy out. If you don't have a meat grinder, set the trim aside and use it in stews and meat sauces or dice it by hand."

Many butchers consider a 80/20 ratio of lean to fat for grinds standard. Although pork grind can sometimes be appreciated for its lean quality (if you grind from the leg or from the leaner shoulder cuts), if you plan to use the grind for sausage-making (refer to Chapter 16), make sure to save some extra fatback to add to your grind.

Part IV
Beef Butchery

The 5th Wave   By Rich Tennant

"You gonna get that cow shelter built soon?"

In this part . . .

*I*n this part, I explain how to butcher beef, probably the largest animal you'll tackle as a home butcher. With inside advice from one of the best butchers around, Oscar Yedra, you'll discover how to butcher a side of beef. I also tell you about beef primals, subprimals, and retail cuts.

Chapter 11

What's Your Beef? Understanding the Cuts

. . . .

In This Chapter

▶ Reviewing general principles of beef butchery

▶ Familiarizing yourself with beef cuts

. . . .

*B*eef is an intimidating beast to butcher. The raw poundage and the financial investment alone are enough to keep novices away and seasoned professionals on their toes. Beef is large, heavy, and cumbersome (a side of beef can weigh up to 800 pounds), and it has a generous amount of fat and dense bones to deal with. If you're not careful, butchering beef can also be dangerous.

Yet beef butchery, when done well and safely, is practically an art form that not only yields a ton (almost literally!) of meat, but also helps to preserve an endangered craft.

In this chapter, I explain the basics of beef butchery and introduce you to the primal, subprimal, and retail beef cuts. Consider this chapter a primer on beef butchery, preparing you for the actual task of butchering a beef forequarter and hindquarter, the subjects of Chapters 12 and 13.

The Lowdown on Beef Butchery

The term "beef" refers to meat from bovines, particularly that of domesticated cattle (which is probably the beef you're most familiar with), bison (which is growing in popularity), water buffalo, and even yak.

Familiar domestic cattle breeds include Angus, Hereford, Shorthorn, Wagyu, and Devon, and the weight of the animal and muscle characteristics change from breed to breed. Regardless of the breed, the instructions are the same.

Preserving tradition

Breaking beef carcasses in a retail setting is a dying art form for a number of reasons, many of them logistical: what to do with all that meat when you're done, for example, or how to offset the cost of the labor and the waste. Making the problem more challenging is finding opportunities to learn how to cut beef when so few opportunities for education exist. (For a detailed discussion of why butchery is endangered — due largely to the centralization of meat processing — refer to Chapter 1.)

These challenges can seem daunting, but beef butchery is a craft that requires strength, knowledge, creativity, and dedication. When a masterful talent like beef breaking is nearly extinguished, we lose not only a skill but also the process itself and every relationship, consideration, and connection within it: connection to farmers, master butchers, and those who value their contributions, and connection to the land and our artisan past. We also lose the knowledge and skill that enable us to prepare more unusual cuts of beef and the talent to seam out muscles.

Skillful and passionate butchers have a responsibility to educate others in order to keep the craft of butchery alive.

Muscles matter! Paying attention to beef musculature

To be proficient at beef butchery, you need to know your meat musculature. Although you don't need to study it with the dedication of an anatomist, you do need to have a basic familiarity with where the muscle groups are and what they do. Why? Because the work of the muscle — the way the animal moved and lived — determines its tenderness. Working muscle is tougher muscle. For example, the muscles that run down either side of the spine don't work as hard as leg muscles. This is why the rib and the loin are more prized, tender grilling cuts, and cuts from the round (leg) are leaner and tougher.

In addition, you want to keep muscle groups intact as much as possible when pulling them off the carcass. If you cut through a muscle group in the wrong place, you impede your ability to get the most steaks or roasts out of it, and you end up with valuable meat in your grind pile.

The most difficult part of beef butchery for the novice or aspiring meat cutter is understanding where the cuts come from, how to produce them, and what they are called. Don't get overwhelmed; it may be daunting, but it is not impossibly out of reach.

To educate yourself on musculature, you can go to a variety of wonderful resources, like *The Art of Beef Cutting,* by Kari Underly, and *The North American Meat Processors Association* (NAMP) *Guide.* You can also practice on smaller animals, which helps you become familiar with seaming out muscle groups and gives you an opportunity to practice proper trimming technique. (The butchery instructions in Part II are ideal.) And don't be afraid to make mistakes.

Maximizing flavor and tenderness

Getting the most out of your beef requires more than just cutting off hunks of meat and frying or grilling it up. In addition to knowing where the different cuts come from, you also need to know how to prepare them to maximize their flavor and tenderness. Here are the things to keep in mind (you can find specific instructions on accomplishing these tasks in the next chapters):

- **Fat is flavor, so whenever possible try to leave some fat on your cuts.** However, in some cases, you need to remove some of the fat cap, either as an aesthetic choice or out of necessity to get at the inside layers of interconnective tissue (these affect the outcome of the meat during cooking).

- **Connective tissue, silver skin, and tendons are tough and hard to chew, but they add rich flavor to dishes — provided you use the right cooking method.** You prepare these kinds of cuts (beef shank and chuck are examples) with slow, moist cooking methods. In the case of roasting or grilling cuts, you want to remove these tissues. (This is another reason why truly great butchers understand cooking: They know what kinds of cuts are best for the cooking method their customers are planning.)

- **Don't waste.** Being wasteful is pointless and disrespectful. Almost everything on an animal can be consumed or used in some way. Wasting meat negatively affects your bottom line, whether you're a hobbyist who simply wants to provide meat for your own family, or a professional butcher who works in a meat shop. To make use of as much of the beef as you can, have a plan before you butcher so that you aren't overloaded with fats and trim, which you may be tempted to throw away.

- **Think about how the final cut will look and be prepared.** Retail cuts should be attractive and produced in a way that is well-suited to how they will be prepared. For example, grilling cuts like rib eye or New York strip are trimmed of fat and cut into individual steaks that are ready to throw on the grill. Braising cuts like pot roast or chuck are left in large pieces so that they can be slow cooked or smoked and portioned afterwards.

- **As a novice, avoid producing bone-in beef cuts.** Sawing through thick bone with a handsaw is difficult; in fact, many experienced butchers use a bandsaw expressly for this purpose. For that reason — and at least while you're still learning the ropes — avoid producing bone-in cuts. Debone the meat.

Playing it safe

Due to the size and weight of a beef carcass, butchering it can be a big challenge, especially for novice butchers. Before you decide to take on this task, make sure you can answer "yes" to each of the following questions:

✔ **Do you know how to use the butchery equipment (knives, saws, cleavers, and so on) competently and safely?** If not, practice some more on small animals. Chapters 5 through 7 give you a chance to tackle smaller butchery jobs. And make sure you know knife safety backward and forward; Chapter 4 has the details.

✔ **Are you diligent about taking safety precautions?** Make sure to keep your knives in a scabbard while cutting so that you don't cut yourself by grabbing a knife you precariously left on the table. You may also choose to wear a chain-mail apron or cut gloves. The point? Don't take safety lightly; pay attention because it's important.

✔ **Are you strong enough (or have a partner who's strong enough) to handle the weight of the carcass?** A hindquarter of beef can weigh 250 or more pounds. Trying to lift it on your own is a bad idea.

✔ **Do you have the time necessary to do the job without rushing?** From start (primals) to finish (retail cuts), butchering beef quarters can take anywhere from 2 to 8 hours, depending on your experience level. It's not a job you can rush and do either well or safely. So make sure you have the sufficient amount of time free to devote to the task without having to take shortcuts.

✔ **Do you have the space necessary to keep the meat safely chilled during the butchery process?** As I mention earlier, butchering a cow takes time, and you're dealing with a lot of meat — meat that you have to keep chilled while you butcher and that you have to have the space to store when you're done. If you are short on space, you could purchase some large ice chests and rotate cuts in and out of refrigeration while you are butchering, but this is not a permanent solution. If you plan to butcher at home, invest in refrigeration space to store the meat when you are done.

Dividing Up the Task: Primals, Subprimals, and Retail Cuts

In the United States, beef is typically first halved and then quartered (how specifically the whole animal is initially divided varies depending on the geographical region, tradition and habits of that area, and demand for high-quality meat). These four parts are then sold as is or cut down into primals and subprimals prior to being boxed up and shipped out. (Head to Chapter 3 for a general explanation of primal and subprimal cuts.)

Each primal, or main section of the carcass, has attributes (such as tenderness, marbling, flavor, and so on) that are unique because of that section's function and role in locomotion when the animal was alive. For example, from the chuck primal, you get cuts that are renowned for being suitable for braising or grinding. The rib is known for its prized grilling cuts, and the round, with its combination of tender and moderately tender, lean muscle groups, is famous for wonderful roasting meats. Brisket and shanks are famous as slow-cooker meats, and flanks for quick grilling and grinding.

Forequarter and hindquarter primals and subprimals

The forequarter is the anterior (the front) portion of the carcass, which includes the chuck, the plate, the brisket and the rib. The hindquarter is the posterior (or back) portion of the carcass and includes the loin, the sirloin, the round, and the flank. Figure 11-1 shows the fore- and hindquarter primals and the subprimals for each section.

The primals on beef, or any other animal for that matter, change from country to country, because each country has its own cooking traditions and customs. In the United States, for example, we like fast preparations and premium meat; as a result, butchers cut as many steaks as possible.

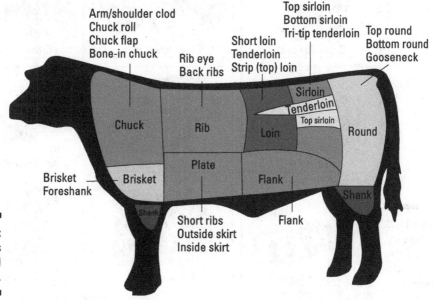

Figure 11-1:
Beef primals and subprimals.

Illustration by Wiley, Composition Services Graphics

The retail cuts

In the following sections, I list a variety of retail cuts that come from the different primal sections of the cow. Keep in mind, however, that this list isn't definitive, nor is it complete. A beef carcass can be fabricated in a number of ways, resulting in different cuts than those listed here. The cuts highlighted here are the ones that I discuss in this book.

Cuts from the chuck

The chuck section comes from the shoulder and neck of the beef, and it yields some of the most flavorful and economical cuts of meat. Figure 11-2 shows cuts from the chuck.

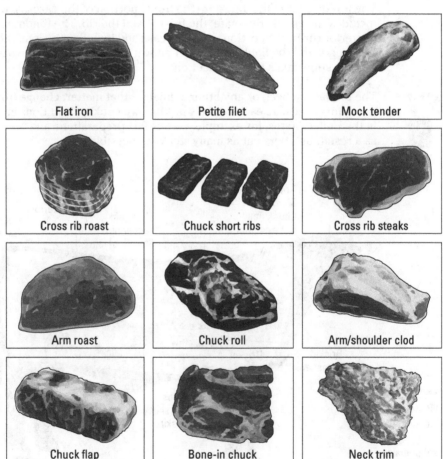

Figure 11-2: Retail cuts from the chuck.

Flat iron · Petite filet · Mock tender
Cross rib roast · Chuck short ribs · Cross rib steaks
Arm roast · Chuck roll · Arm/shoulder clod
Chuck flap · Bone-in chuck · Neck trim

Illustration by Wiley, Composition Services Graphics

Cuts from the plate

The plate (also known as the short plate) is from the front belly of the cow, just below the rib cut. The short plate produces types of steak such as short ribs and skirt steak. These and other cuts are shown in Figure 11-3.

Outside skirt

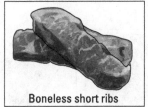
Short ribs

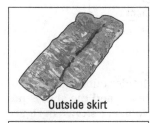
Inside skirt

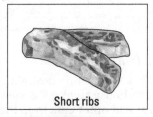
Boneless short ribs

Figure 11-3: Retail cuts from the plate.

Illustration by Wiley, Composition Services Graphics

Cuts from the rib

The rib is known for prized cuts such as the rib eye (bone-in and boneless) and export rib, which is used for prime rib. You can also cut exceptional steaks from the rib eye cap. These cuts are shown in Figure 11-4.

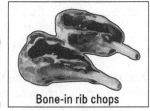
Bone-in rib chops

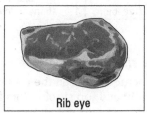
Rib eye

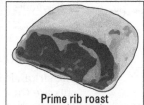
Prime rib roast

Figure 11-4: Retail cuts from the rib.

Illustration by Wiley, Composition Services Graphics

Cuts from the brisket and foreshank

Brisket is a cut of meat from the cow's breast or lower chest. The muscles in the brisket support most of the animal's body weight when it's standing or moving. The point? Cuts from this area have lots of connective tissue and therefore require long, slow cooking (think slow cooker or low, slow roasts) to help tenderize them. Cuts from the brisket and foreshank are shown in Figure 11-5.

Figure 11-5:
Retail cuts
from the rib.

Brisket
Cross-cut shanks

Illustration by Wiley, Composition Services Graphics

Cuts from the loin or short loin

The short loin is known for its desirable, tender cuts of meat such as the tenderloin, porterhouse steaks, T-bone steaks, and the always popular strip loin. These cuts, and other cuts from the loin, are shown in Figure 11-6.

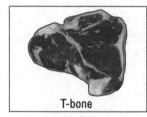

T-bone
Porterhouse steak
Filets

Figure 11-6:
Retail cuts
from the loin
or short loin.

New York strip (top loin)
Club steaks
(bone-in loin steaks)
Hanger steaks

Illustration by Wiley, Composition Services Graphics

Cuts from the sirloin

The sirloin is located between the round and the short loin. Some familiar cuts are the tri-tip and top sirloin. Cuts from the sirloin are great for roasting. These and other cuts cuts are shown in Figure 11-7.

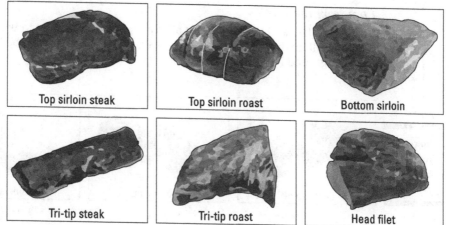

Figure 11-7: Retail cuts from the sirloin.

Illustration by Wiley, Composition Services Graphics

Cuts from the round

The beef round is the back leg of the animal. The cuts from the round are leaner and tougher due to the fact that these muscles work hard. Cuts within the round are best prepared using moist heat. These cuts are shown in Figure 11-8.

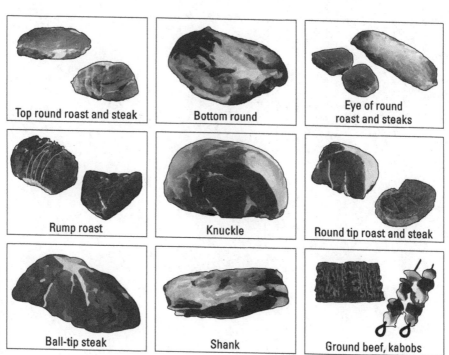

Figure 11-8: Retail cuts from the round.

Illustration by Wiley, Composition Services Graphics

Cuts from the flank

Beef flank, flap, and skirt steak are all great grilling cuts located within the flank (if grilling them, go for medium rare). These cuts can also be braised or used for ground beef. They're shown in Figure 11-9.

Figure 11-9:
Retail cuts from the flank.

Flank steak

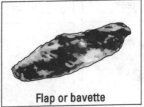

Flap or bavette

Inside skirt

Illustration by Wiley, Composition Services Graphics

Chapter 12

Beef: The Forequarter

. . . .

In This Chapter

▶ Cutting the forequarter into primals and subprimals

▶ Getting a variety of retail cuts

▶ Making use of the "waste"

. . . .

The *forequarter* is the front half of a side of beef. It's made up of the fore-limb; one half of the chest, including ribs; and the neck. All four-legged livestock have the same makeup in their forequarter, but generally beef is the only one large enough to necessitate being divided in this way. In this chapter, you discover how to break the forequarter into primals (the first, or primary, portions that the carcass is broken into), then subprimals (cuts larger than steaks or roasts, typically isolated muscles), and finally into retail cuts.

Breaking the Forequarter: The Basics

A forequarter of beef has four primals: the chuck, the plate, the rib, and the brisket/foreshank. Each primal can be cut down into subprimals that are themselves then cut down into the retail cuts you're probably already familiar with. Table 12-1 lists these different cuts. Keep in mind that this table doesn't include all the possible retail cuts (especially in the chuck). But it does give you a general idea of how the work flows and the cuts you'll be able to produce. *Note:* The notations "on the rail" or "on the bench" indicate where the cut is made. You make "on the rail" cuts with the forequarter hanging on the rail; for cuts made "on the bench," you move the meat to a butcher block or work surface to finish the cutting.

Table 12-1 Primal, Subprimal, and Retail Cuts from the Forequarter

Primals (on the Rail)	Subprimals (on the Bench)	Retail Cuts (on the Bench)
Chuck	Arm/shoulder clod, chuck roll, chuck flap, bone-in chuck	Flat iron, petite filet, mock tender, cross-rib roast, cross-rib steaks, arm roast, chuck roll
Plate	Short ribs, outside skirt, inside skirt	Outside skirt, short ribs, inside skirt, boneless short ribs
Rib	Ribeye, back ribs (export or roll)	Bone-in rib chops
Brisket/foreshank	Brisket, foreshank	Brisket cross-cut shanks

From each section of the animal, you also end up with trim that can be used for grind (ground beef), stew meat, stocks, and sausage-making.

Fashioning a hook and rail

You can break down a side or quarter of beef in many different ways. In this chapter, I explain how to break beef while the forequarter hangs from a hook and rail. This method lets you use gravity to make the job easier and gives you better accessibility. It's pretty safe to assume, though, that most people don't have the equipment to hang a quarter of beef at home.

Fear not!

In meat processing companies, a hook and rail system involves a series of rails that hang from the ceiling and that are used to move and hang carcasses. Trolley hooks anchored to the rail are used to move the hanging carcass back and forth between the refrigeration area and the work area so that the butcher can cut the meat as it hangs — nice and efficient, no? Although a hook and rail system may be standard equipment in large meat processing companies, these meat moving systems are very rare; some old school butcher shops may be lucky enough to have them, but that's generally not the case.

So how can you set yourself up to hang meat at home? I have been to many workshops, classes, and events where we have had to hang carcasses without a hook and rail system. Here's what I recommend: Find a support beam in the ceiling of your garage or shed that can handle the weight of the meat. Anchor a long length of strong chain around the support beam (you can use a climbing hook to link the chain together). Then anchor a large S hook through the chain (where the climbing hook is), and hang the carcass from the hook.

Alternatively, if you don't have a support beam that can handle the weight, you can build an A-frame (see Figure 12-1) and use that in the same manner.

Figure 12-1:
An A-frame
hook and
rail set up.

Illustration by Wiley, Composition Services Graphics

Cutting on the rail

When hanging a forequarter, anchor the hook in between the first few ribs with the arm hanging down (see Figure 12-2). Be careful not to pierce the skirt steak; doing so damages the cut. When hanging a hindquarter, also shown in Figure 12-2, hook the leg through the Achilles tendon, which is strong enough to support the weight of the carcass.

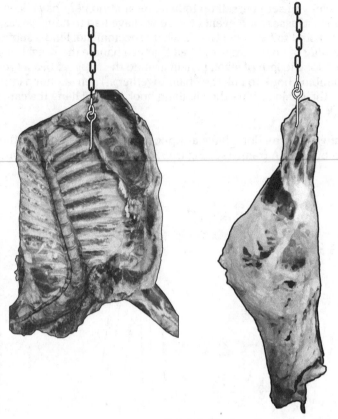

Figure 12-2:
Hanging the
forequarter
between the
first few ribs
(left) and the
hindquarter
from the
Achilles ten-
don (right).

Illustration by Wiley, Composition Services Graphics

When it comes to safety and butchering a forequarter of beef, you have a couple of things to think about: your own safety and keeping the meat cold while you work. Here are the details:

✔ **Wear the appropriate clothing and safety gear.** This includes good non-slip shoes. You may also choose to wear a chain-mail apron, cut gloves, and a scabbard and chain to keep your knives safely at your hip instead of resting precariously on the table. Refer to Chapter 4 for general safety advice and information on using knives safely.

✔ **Keep the beef cool.** Keep the quarter or side of beef cold at all times. When you're ready to begin cutting, take the quarter out to your work station, break it quickly, and then put the pieces back in the refrigerator as they come off the carcass. Be sure to refrigerate the subprimals until you're ready to do the trim work.

Removing the Outside Skirt (Rail)

The outside skirt, also known as the diaphragm, is a well-loved cut of meat because it is very flavorful and is wonderful grilled or quickly seared medium rare.

The outside skirt is located on the underside of the plate. Because it hangs off the body, it's the first thing you pull off. Anchored to the ribs between rib 6 and rib 12, the boneless skirt hangs off the ribs, much like a flap of meat, making it easy to identify (see Figure 12-3). The skirt is encased in a tough (and usually dried) piece of membrane and must be trimmed before being prepared.

To remove and prepare the outside skirt, follow these steps, shown in Figure 12-3:

1. **Hold your boning knife in a butcher's grip and score down the skirt (on top of the bone), just along where the skirt attaches to the ribs (Figure 12-3a).**

   This score mark is the line you'll follow in Step 2. For knife grips, refer to Chapter 4.

2. **Grab the top of the outside skirt with your free hand and, while pulling up on the skirt as you cut it free, slice close to the bone at the base of the skirt steak (where you scored it in the Step 1).**

3. **Continue pulling the skirt and slicing it away from the rib bones down the entire length of the outside skirt (Figure 12-3b) until it is removed.**

   With the outside skirt removed from the forequarter, you need to "unlock" it by removing a chain of connective tissue that runs along the perimeter of the outside skirt.

4. **Lay the skirt flat on the cutting surface in front of you. Then slice 1/2- to 3/4-inch off the entire length of the skirt (Figure 12-3c).**

   From Oscar Yedra, Canyon Market, BN Ranch: "By 'unlocking,' or cutting the thick connective tissue that runs along the side of the skirt, you will easily be able to pull off the membrane with your hands."

5. **Rotate the outside skirt so that the section you just cut is facing you. Put your knife down and use your hands and fingers to pull the membrane away from the muscle on both sides of the skirt (Figure 12-3d).**

   The outside skirt is now ready to sell or prepare.

   One attractive way to present the outside skirt is to roll it and put it on a papered tray for display.

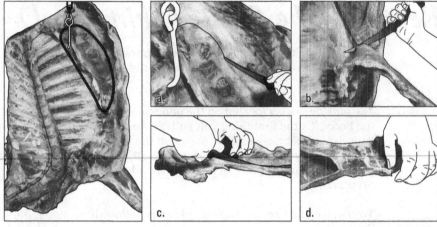

Figure 12-3:
Removing
the outside
skirt.

Separating Out the Chuck, Arm, and Brisket from the Plate and Rib (Rail)

In this section, you produce several subprimals through a series of purposeful cuts that are intended to accomplish a few things. First, you score around select muscle groups to make the cuts easy to remove. Second, you make marking cuts that divide and section the primals. Last, you cut the supbrimals free, using a meat hook, a breaking knife, and a boning knife. When you're done, you'll have a chuck, an arm/shoulder, and a brisket.

Here's the order in which you accomplish these tasks:

1. Identify and mark the point where the chuck is separated from the plate and rib (or export rib).

2. Cut through the muscle where the chuck is separated from the plate and rib.

3. Score around the brisket.

4. Remove the arm from the chuck.

5. Remove the brisket as one complete muscle, preserving a larger portion than may be typical.

6. Remove the neck meat and atlas joint.

7. Remove the flat iron.

8. Saw the chuck free from the plate and rib.

Figure 12-4 shows where you'll make these cuts.

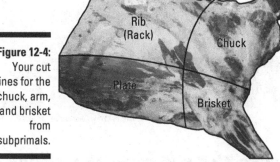

Figure 12-4:
Your cut
lines for the
chuck, arm,
and brisket
from
subprimals.

Illustration by Wiley, Composition Services Graphics

At this stage in breaking, you are sectioning the forequarter into manageable pieces. More recognizable retail cuts will begin to emerge in the later stages of cutting.

A little prep work (scoring around muscle groups and marking your cuts, tasks you do in the following sections) goes a long way. Not only does it make the task easier when the time comes to remove sections of the carcass (which are heavy, cumbersome, and difficult to maneuver), but it also ensures that you are cutting into the right muscle groups — or more importantly, not cutting into the wrong ones. Butcher's Guild Member Oscar Yedra shared this technique with me, and it's the method you'll use frequently when you're butchering a forequarter of beef.

Step 1: Marking the chuck and rib

The arm chuck/brisket section has five ribs on it; the plate/rib section has seven. So the first thing you need to do is to count the ribs and make a cut mark as shown in Figure 12-5:

1. **Begin by counting up five ribs, starting at the first rib (the one anterior of the spine). Mark the location above the fifth rib by pushing your knife through the meat to the other side (Figure 12-5a through c).**

   Push your knife about four inches away from the spine into the muscle.

2. **Flip the forequarter around so that the outside is facing you. You can see where your knife pushed through to the other side.**

Figure 12-5:
Marking the
chuck and
rib.

Illustration by Wiley, Composition Services Graphics

Step 2: Separating the rib from the chuck

Starting at the cut you made from the inside of the forequarter, use your breaking knife to cut through the meat on the outside of the forequarter in a reasonably straight line parallel to the ribs, separating the rib from the chuck (see Figure 12-6). Stop at the spine.

Figure 12-6:
Separating
the rib from
the chuck.

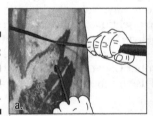 

Illustration by Wiley, Composition Services Graphics

You will never be able to cut through bone with a knife, so don't even try. It's dangerous. What you are trying to do is cut through the muscle so that when you are ready to saw through the spine, you won't damage the flesh in the process.

Step 3: Scoring the brisket

Some butchers cut through the brisket by continuing the trajectory of the cut they make when they mark the chuck and rib (as explained in the preceding section) all the way to the edge of the breastbone. The technique I show you here differs in that, instead of cutting straight through the brisket, you make a scoring cut in the approximate shape of a half-moon. For this cut, you start approximately 4 inches above the original marking cut, starting at the sternum. Scoring the brisket this way allows you to extract a larger portion of brisket. (The meat from this section is also used for navel pastrami.)

To score around the brisket, follow these steps:

1. **Start at the breastbone, or sternum, about 4 inches above the cut you made to mark the chuck/rib on the outside of the carcass.**

2. **Slice along the top of the muscle and curve down with your knife to reach the point where the arm connects to the brisket, ending just shy of your "marking the chuck/rib" cut (see Figure 12-7).**

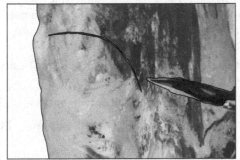

Figure 12-7:
Scoring the
brisket

Illustration by Wiley, Composition Services Graphics

Step 4: Removing the arm from the chuck

In this step, you pull the arm from the chuck while the forequarter is still hanging on the rail. Follow these steps, shown in Figure 12-8:

1. **Look inside the cut marking the chuck/rib to find the shoulder blade (also called the *scapula* or the *blade bone*).**

2. **Anchor your boning knife at a 45 degree angle off the exposed blade bone. Then use a butcher's grip and pull down, letting your knife ride against the bone until you hit the arm bone (where the joint connects the arm to the shoulder bone) (Figure 12-8a).**

3. **When you hit the arm bone, use your knife to whip around the ball joint, in-between the socket (Figure 12-8b).**

This cut is a little tricky, but do your best. Don't get discouraged! Becoming an expert butcher takes a lot of practice.

4. **Start with the top corner of the arm where your "marking the chuck/rib" cut and your "brisket scoring cut" meet. With your boning knife, slice the meat free as you pull down with your hook, using gravity to assist you (Figure 12-8c).**

As your knife moves down the arm subprimal, you need to reposition your hook to gain better control of the large portion of muscle. As you cut, keep moving the hook in closer to where you're cutting so that you can use it to pull the meat away from the interconnective tissue while simultaneously cutting through the flesh.

After you cut most of the arm free from the chuck, you'll find that it's still attached at the arm joint. Because you made that tricky cut — the one where you whipped around the joint in Step 3 — you don't have to struggle through the meat to find where to cut. The ball and socket is exposed (Figure 12-8d) and you can easily cut through it.

5. **Cut through the ball and socket to sever the arm joint (Figure 12-8e and f).**

Now the arm joint is attached to the chuck only by muscle in the armpit.

6. **Use your hook to grab the arm and cut it completely free of the carcass (Figure 12-8 g).**

7. **Set the arm/shoulder clod aside, under refrigeration, until you can process it into retail cuts.**

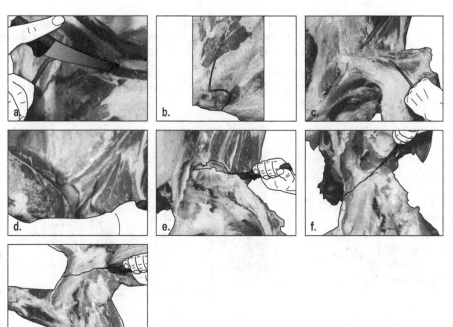

Figure 12-8:
Removing
the arm
from the
chuck.

Illustration by Wiley, Composition Services Graphics

Step 5: Removing the brisket

To remove the brisket, follow these steps:

1. **Score around the side of the brisket against the sternum: Holding your boning knife in a butcher's grip, use the knife tip to cut along the bone until you reach the bottom (Figure 12-9a).**

   Let your knife ride against the sternum, as you cut down. Scoring around the bones and the muscle groups will assist you when you're ready to pull off the entire piece.

2. **When you reach the bottom of the sternum, make a straight cut down (parallel to the spine) at the base of the neck through the meat (Figure 12-9b).**

   Now you're ready to cut the brisket free.

3. **Use your hook to grab the top corner of the brisket and pull it down. Then slice through the seam, or connective tissue, while simultaneously pulling down with your hook (Figure 12-9c).**

   The muscle will begin come away from the forequarter pretty quickly so keep cutting from left to right across the underside of the brisket until it is removed.

Figure 12-9:
Removing
the brisket.

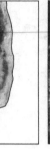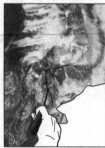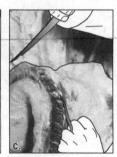

Illustration by Wiley, Composition Services Graphics

Step 6: Removing the neck meat and atlas joint

The atlas joint is the topmost vertebra, where the skull attaches to the spine. You can identify the atlas joint by simply finding the first vertebra at the top of the spine. Removing the atlas joint along with the neck meat lets you remove the neck bones more easily later (this joint essentially "locks" the vertebra in place).

To remove the neck meat and atlas joint, follow these steps (see Figure 12-10):

1. **Locate the first vertebra at the top of the spine; this is the atlas joint (Figure 12-10a).**

2. **Slice down the spine, starting at the edge where the brisket was removed, and cut the neck meat away from the carcass until you reach the atlas joint (Figure 12-10b).**

3. **Cut between the vertebrae and remove the atlas joint, leaving it attached to the neck meat (Figure 12-10c).**

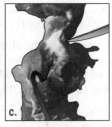

Figure 12-10:
Removing
the neck
meat and
atlas joint.

a.   b.   c.

Illustration by Wiley, Composition Services Graphics

Neck meat can be stewed, ground, or even tied into a roast and then cut into portion-sized medallions for a nice presentation in the meat case or when braising at home.

Step 7: Removing the flat iron

Removing the flat iron (also called the *top blade steak*) on the rail allows you to use the stability and weight of the forequarter to help you yank the flat iron off in one piece. Doing so leaves you with a clean shoulder blade — and an impressive feeling of accomplishment.

Ah yes, ripping the flat iron is quite a satisfying feat of butchery, if you can get it right. It requires that you score completely around the muscle first so that, when you grab the muscle with your hook and pull, it comes right off. Then the crowd roars.

To remove the flat iron, follow these steps, shown in Figure 12-11:

1. **Feel the end of the shoulder blade (the blade bone), where it was attached to the arm bone (Figure 12-11a).**

   You want to begin scoring around the flat iron at the base of this socket.

2. **Insert your boning knife at the base of the shoulder blade and cut up along the shoulder blade, ending where you sectioned the rib from the plate (Figure 12-11b).**

   Remember, you haven't removed the rib from the plate yet, but you made a cut sectioning it from the plate (refer to the earlier section "Step 2: Separating the rib from the chuck").

3. **Cut all the way up the flat iron, following the blade bone all the way up to the exposed, cut top edge of the shoulder blade (Figure 12-11c).**

   Use your hook to steady the forequarter as you cut.

Now that the flat iron, or top blade steak, has been scored, you can "rip" it from the top of the shoulder blade.

4. **Use your meat hook to grab the bottom corner of the flat iron at the base of the shoulder blade. With your boning knife, cut the meat on top of the shoulder blade free, going all the way through the membrane to the bone (Figure 12-11d).**

   This is your starting point and where you will anchor your hook to pull the muscle off.

   In spirit, this step is like peeling the corner of a sticker back to get it "started." Sticking your thumbnail into the sticker's edge and peeling it back with your other hand is intuitive, and helps it to come off in one clean piece.

5. **Put your knife down and, with the meat hook in your dominant hand, hook the end of the flat iron. (Make sure the hook is grabbing under the membrane.) Steady your free hand against the base of the shoulder blade to create some resistance and yank the flat iron back (quick and strong), pulling it off of the shoulder blade in one piece (Figure 12-11e).**

6. **Set the flat iron aside, under refrigeration, until it is ready to be trimmed.**

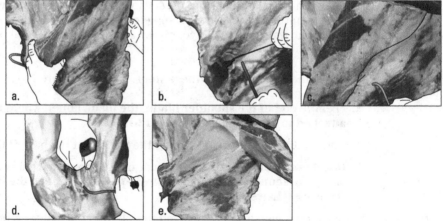

Figure 12-11: Removing the flat iron.

Illustration by Wiley, Composition Services Graphics

Step 8: Removing the chuck

You have already done the necessary knife work to separate the chuck and the rib. At this point you are ready to remove the chuck. Follow these steps, shown in Figure 12-12:

1. **Saw through the spine (Figure 12-12a).**

   Now you're ready to saw through the ribs.

2. **On the inside of the forequarter, score a line down the ribs, starting approximately 5 inches from the spine on the fifth rib (parallel to the sternum) (Figure 12-12b).**

   The amount of bone on each rib within the chuck doesn't have to be even. You'll remove these bones on the bench, or cutting table, so make sure your cut line follows the angle of the sternum, not the spine. Doing so produces an attractive, square section of chuck short ribs when you start making the retail cuts.

3. **Saw through the ribs, following your cut line (Figure 12-12c).**

4. **Switch to your breaking knife or cimeter and cut through any remaining meat.**

   Make sure your meat hook is anchored to the chuck. As soon as you cut the chuck free, it will fall off the carcass fast and hard, and you need to catch it.

   Be prepared to absorb the weight of the drop with your arm. In addition to holding the hooked chuck with your free hand, you can also position you leg under the chuck to help absorb the drop. (A whole chuck can weigh up to 100 pounds, or an estimated 25 percent of the whole carcass weight. This chuck will weigh less because you've already removed the arm/shoulder clod and the flat iron.)

Figure 12-12: Removing the chuck.

Illustration by Wiley, Composition Services Graphics

Squaring Up the Chuck Short Ribs (Rail)

The chuck short ribs come from the remaining section of bones under the brisket. They are fatty, delicious, rich, portions of meat that can be sold and served with or without the bone. (I prefer to leave the bones in for added flavor and presentation.) In addition to braising or slow roasting, you can use the meat from the chuck short ribs for grinding into sausage or as boneless stew meat. Make sure to set all bones aside for the stock pot.

To square up the short ribs, follow these steps:

1. **Take a look at the plate and ribs and identify where the sternum ends near the neck (Figure 12-13a).**

    The sternum is the very thick section of bone on which the point of the brisket laid.

    You want to produce a rectangular, even section of chuck short ribs. To do this, you need to make sure your cut line starts at the top of the chuck ribs and ends at the base of the sternum.

2. **Drag your knife down the ribs where you intend to saw to create a guideline to follow (Figure 12-13b).**

3. **Hook the chuck ribs with your meat hook, steadying the rib section as you saw through the chuck (Figure 12-13c).**

    You will be left with a small section of brisket bone hanging off the rib and plate.

4. **Saw the remaining section of brisket bone, or sternum, from the hind-quarter (Figure 12-13d).**

    Set this aside for grind or the stockpot.

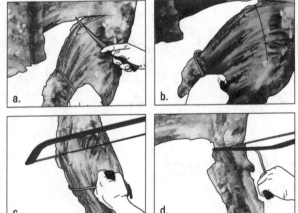

Figure 12-13: Squaring up the chuck short ribs.

Illustration by Wiley, Composition Services Graphics

Sectioning the Rib from the Plate (Rail)

The rib is known for prized cuts such as the rib eye, bone-in rib eye, and export rib (used for prime rib). You can also cut exceptional steaks from the rib eye cap. The plate is used for grinding, stew meat, or short ribs.

To section the rib from the plate, follow these steps, shown in Figure 12-14:

1. **Mark the rib (top) where your cut line starts (Figure 12-14a).**

   Because the meat on the export rib is so valuable, don't cut shy of the plate and into the rib meat. Use either of these two methods to determine where your cut line should start:

   • Look at the loin muscle at the top of the rib. Notice where it ends with a triangular-shaped piece of fat connecting to the ribs. This is where you make your cut.

   • Measure the distance from thumb to pinky (thumb touching the spine, pinky touching the ribs) on the top of the rib-loin end. The tip of your pinky is where you start your cut.

2. **Mark the plate (bottom) where your cut ends (Figure 12-14b).**

   At the base of the ribs, measure four fingers width from the spine. At the edge of your index finger is where the cut ends. Make a notch in the meat.

3. **Drag your knife over the ribs, starting at the notch located at the top of the rib, or plate, and joining it with the notch at the base of the plate.**

4. **Following your cut line, saw through the ribs while holding the rib steady with your free hand (Figure 12-14c).**

   As you saw down through the ribs, be sure to hook the rib with your meat hook and be prepared for the drop.

5. **Set the export rib (Figure 12-14d) aside, under refrigeration, until you're ready to cut further.**

   You can store the export rib in the refrigerator for up to two weeks before cutting, if you want to age it.

6. **Remove the remaining plate section from the hook and set it aside, under refrigeration, until you're ready to make retail cuts.**

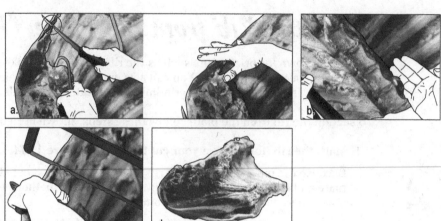

Illustration by Wiley, Composition Services Graphics

Figure 12-14:
Sectioning
the rib from
the plate.

Trimming the Brisket (Bench)

After you section the forequarter into its primals and subprimals, you're ready to start making your retail cuts. First up is the brisket.

Brisket can be served braised, smoked, brined, or stewed. It's a well-known barbecue favorite and can be sold fresh in a butcher's case as a whole brisket, cubed for stew meat, ground up for burger blends, or transformed into corned beef through brining and cooking.

Essentially, when you trim the brisket, you remove some fat from it. You may choose to trim about 1/2 inch off the top, but the main goal is to give the brisket a nice shape and even out the amount of fat on the surface of the cut.

Follow these steps, shown in Figure 12-15:

1. **Lay the brisket on the cutting board and use a large butcher knife or cimeter to cut the fat away from the brisket (Figure 12-15a).**

   Always make sure to cut away from your body.

2. **Round off the corners and square up the edges of the brisket on the side you are trimming (Figure 12-15b).**

3. **Flip the brisket over and trim a bit off the surface of the point or any neck meat (Figure 12-15c).**

4. **Trim off any yellow, hardened fat that was exposed to air (during aging or in transport) to expose the white fat underneath. Then trim to the desired thickness (Figure 12-15d).**

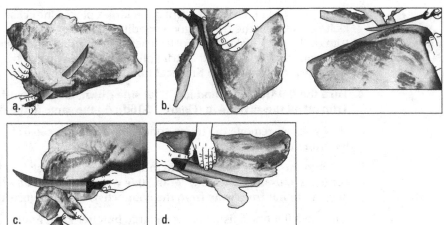

Figure 12-15:
Trimming
the brisket.

Illustration by Wiley, Composition Services Graphics

Trimming the Flat Iron (Bench)

The flat iron has become hugely popular over the last decade due to its nice shape and semi-marbled, yet generally lean, tender muscle profile. The flat iron has one major flaw though: Running right through its center is a very tough, inedible, thick layer of connective tissue. If you remove the connective tissue, however, you end up with two lovely steaks. You can grill or quickly sear this cut medium rare and slice it against the grain to serve.

To trim the flat iron, follow these steps, shown in Figure 12-16:

1. **Lay the flat iron, fat side down, on the cutting surface in front of you.**

2. **Using a large butcher's knife or cimeter, trim off some of the meat on the edge of the flat iron where the muscle was attached to the shoulder blade, near the socket; continue to trim until you reach the thick layer of connective tissue in the center of the flat iron (Figure 12-16a and b).**

3. **Flip the flat iron over and trim in the same manner on the opposite side, removing some of the lifter meat (Figure 12-16c), again until you reach the outside layer of thick connective tissue.**

   Lifter meat is a term used to describe blade meat that you remove from the shoulder blade. The same term also describes meat from the cap of the export rib. But don't get confused by the two descriptions; they are both part of the same muscle group.

Butchery is as much about cutting the larger pieces into case-ready cuts as it is reserving the trim in good condition for later use. A good butcher trims as he goes, making piles of fat, useable trim, grind, and bones. So make sure you trim for the grind or trim pile as you go. You can trim the fat off lifter meat and then cube the meat for stew meat or to use in sausage.

4. **Turn the flat iron over and lay it, fat side down, on your cutting surface. Trim off all the silver skin (Figure 12-16d); do the same to the other side.**

   Now you can remove the tough connective tissue to separate the whole flat iron into two steaks.

5. **Cut into the center of the muscle to find where the thick inner layer of connective tissue is located; "grab" this with your hook and filet the top part of the flat iron away from the connective tissue (Figure 12-16e).**

   Much like filleting a fish, you use a large butcher knife to filet the top part of the flat iron.

6. **Trim any meat that was left on the surface of the connective tissue by scraping the connective tissue with your knife; set it aside for grind (Figure 12-16f).**

7. **Now turn the flat iron over with the thick connective tissue facing down on the cutting surface and "grab" it with your hook (Figure 12-16g) and filet the remaining portion of flat iron steak by sliding your knife on top of the connective tissue (Figure 12-16h).**

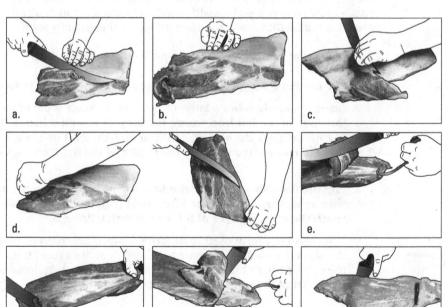

Figure 12-16: Removing the connective tissue from the flat iron.

Illustration by Wiley, Composition Services Graphics

Removing the Foreshank (Bench)

You use the forseshank primarily for stew meat, grind, or osso bucco, which is shank cross-cut slices. These slices can be braised in whole pieces and are big enough to serve two people, depending on how thick you cut them.

Cutting the foreshank from the arm

To remove the foreshank from the arm, you need a breaking knife and a saw. Then follow these steps, shown in Figure 12-17:

1. **Take a look at the shank where it attaches to the arm, near the knee. Pay attention to the yellow circle of fat right above where the fore-shank bones attach to the arm bone (Figure 12-17a).**

2. **Imagine that you are dividing the circle of fat into thirds. Measure in one third of the "circle" starting from the shank side. Make your first cut here (Figure 12-17b).**

3. **Make a slice through the meat and around the bone (Figure 12-17c).**

4. **After you cut through all the meat on the top of the foreshank, switch to a bonesaw and saw through the bone, removing the foreshank from the arm (Figure 12-17d).**

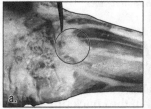
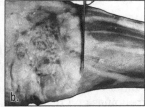
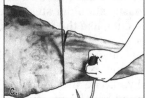
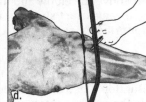

Figure 12-17:
Cutting the foreshank from the arm.

Illustration by Wiley, Composition Services Graphics

Osso bucco

Cutting osso bucco, or crosscut shanks, can be a little tricky with a handsaw, but it can be done. Just make sure the portions are fairly thick so that you don't struggle with the saw too much and tear up the meat in the process.

1. **Slice through the shank meat approximately 1 to 2 inches thick, starting from the knee side (Figure 12-18a).**

2. **Saw through the bone with your bonesaw (Figure 12-18b).**

   Use your free hand to steady the meat so that it doesn't roll while as you saw.

3. **When you are finished sawing, scrape or wipe the meat near the sawed bones.**

   The sawing motion creates friction and forms a substance of emulsified fat, blood, and bone fragments on the meat that decreases shelf life.

Figure 12-18: Cutting osso bucco.

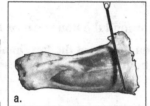
a.

b.

Illustration by Wiley, Composition Services Graphics

The Arm/Shoulder Clod (Bench)

The arm, or shoulder clod, is a subprimal of beef that contains several different muscle groupings and resides within the chuck primal. This muscle system is located above the elbow and mid-blade bone area on the belly side. From this area you can extract shoulder roasts, shoulder steaks, country-style ribs, beef for grind, or smaller muscle cuts, like the petite filet.

Although the chuck is known for producing some of the most delicious slow roasting and braising meats on a cow, it is also the source of many quality burger blends. But don't let the chuck fool you. The shoulder clod brings more to the table than simple stews or ground beef for meatloaf. Grilling and quick-cooking cuts are also within the chuck and shoulder clod. Good marbling and interconnective muscle tissue make for good flavor.

Removing the arm bone

To get to the many delectable cuts housed in the arm/shoulder clod, you first must remove the arm bone. Follow these steps, shown in Figure 12-19:

1. **Lay the arm on the cutting surface in front of you, inside up, and identify the arm bone (Figure 12-19a).**

2. **Use your hook to pull the meat away from the bone as you score around it with your boning knife (Figure 12-19b).**

3. **Continue cutting in this manner, sliding your knife against the bone while pulling the meat away. Work from the top of the bone to the bottom, rolling the bone away from the meat as you cut and slicing and running the back of your knife around the bone on all sides (Figure 12-19c).**

   As you cut around the bone, you expose the arm bone, making it easier to see and work around.

One mark of a good butcher is a clean bone pile. Meat is expensive, and professional butchers don't want to waste precious pennies by leaving good meat on the bone. One way to get clean bones when deboning is to scrape the bone from top to bottom with the back of your knife blade between cuts. This loosens the membranes attached to the bone, helping the meat come "clean off the bone."

4. **Cut the bone free (Figure 12-19d).**

5. **Clean any blood, fat, sinew or connective tissue from the surface of the meat (Figure 12-19e).**

   Use your boning knife to trim the area underneath where the bone rested, which will have some extra fat and blood around it.

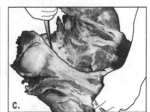
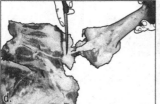
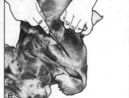

Figure 12-19: Removing the arm bone.

Now that the arm, or shoulder clod, is boneless, you can extract various cuts, depending on your final goals for the meat.

Extracting the petite filet

The petite filet, also known as the *mock tender* or *shoulder tender,* is an oblong muscle, called the *teres major,* located on the inside of the shoulder clod.

To extract the petite filet, follow these steps, shown in Figure 12-20:

1. **Locate the teres major (Figure 12-20a).**

   This muscle lies on top of the largest portion of shoulder at the curve of the arm.

2. **Trim a little fat from the surface to find the seam (Figure 12-20b).**

3. **Remove the muscle in one piece by slicing along the seam while pulling the muscle away from the shoulder (Figure 12-20c).**

   The muscle has some trim, fat, and silver skin attached to it that needs to be removed before it is ready.

4. **Lay the teres major on the cutting board in front of you, outside up. Trim around the sides of the muscle, removing any superfluous meat on either side to isolate the petite filet (Figure 12-20d).**

5. **Trim any silver skin, fat, or membrane from the outside of the petite filet (Figure 12-20e).**

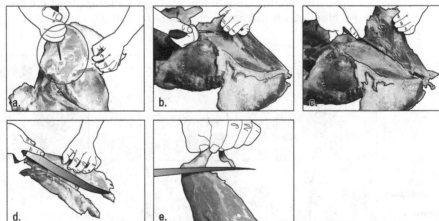

Figure 12-20:
Extracting a
petite filet.

Illustration by Wiley, Composition Services Graphics

Preparing a cross rib roast

The cross rib roast is best prepared by slow roasting under moist heat. It comes from a large section of muscles on the thicker side of the arm/shoulder clod. After you remove the muscle, you can trim it and tie it as a whole roast.

To get a cross rib roast, follow these steps, shown in Figure 12-21:

1. **Trim any superfluous fat or cartilage from the larger grouping of muscles on the arm/shoulder clod (Figure 12-21a).**

2. **Find the seam separating this muscle group from the arm (Figure 12-21b).**

   This is your cross rib roast.

3. **Slice along the seam, pulling with the hook and separating the muscles from each other as you cut. Follow the seam to remove the shoulder clod roast in one large piece (Figures 12-21c and d).**

4. **Trim the fat on the outside of the shoulder clod roast to desired thickness, tie the roast, and voila!**

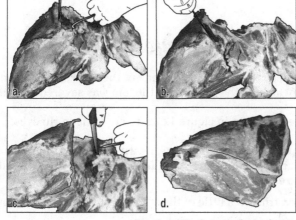

Figure 12-21: Preparing a cross rib roast.

Illustration by Wiley, Composition Services Graphics

You now have one large cross rib roast. But you don't need to stop there. This cut offers you a lot of variety. You can divide the cross rib roast into two portions. From one half, you can slice 1-inch thick cross rib steaks. You can tie the other half into a smaller, family-sized roast.

Tying the arm roast

The final piece of the arm can be trimmed and tied into an arm roast for braising and other slow cooking methods. Simply trim any cartilage, fat, sinew, or thick connective tissue from the arm. When you are satisfied with your trim work, turn the arm over, shape it into a roast, and then tie it.

The Rib and Bone-in Ribeye Steaks (Bench)

Whether you age your export rib or cut steaks right away, bone-in rib chops are a hot item on the home grill or in any meat case. They are well marbled, and when grilled medium-rare and served on the bone, they're juicy, flavorful, and tender. Even the fat is stupendous. The export rib is equally as glorious when prepared boneless and roasted whole for prime rib.

The instructions in this section are for cutting cowboy steaks, also known as bone-in rib eyes. You need a butcher knife and a bonesaw.

Cutting bone-in rib eyes

If your beef has been aged, before you can cut the steaks, trim off any dark crust or dried meat from the side where the loin was severed from the rib.

Follow these steps, shown in Figure 12-22, to cut bone-in rib eyes:

1. **Set the export rib spine, or** *chine,* **down on the cutting surface in front of you. Slice the dried, dark meat from the side of the rib until you hit bone (Figure 12-22a).**

   If necessary, trim off a little of the spine with your bonesaw (or a cleaver and a mallet).

2. **With a large butcher knife or cimeter, slice through the meat between the ribs, determining the desired portion of your first chop (Figure 12-22b). Cut all the way down and through the rib muscle until you reach the chine.**

3. **Saw through the spine (Figure 12-22c).**

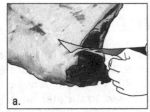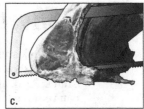

Figure 12-22: Cutting bone-in rib eyes.

Illustration by Wiley, Composition Services Graphics

Frenching the bone-in rib eye

For a more dramatic presentation, impress your friends and neighbors with your crafty cuts and french the bone-in rib chop. *Frenching* is the technique of removing meat to expose the end of a bone, as with a crown roast or rack of lamb. Follow these steps, shown in Figure 12-23:

1. **Hold the bone-in steak with your free hand and rest the rib bone on the table in front of you, meat side up. Cut about halfway down the bone and into the meat (Figure 12-23a).**

   Don't cut too far down; leave a little of that tasty fat and meat on the bones!

2. **Turn your knife blade away from you and scrape down the bone, removing some of the rib meat (Figure 12-23b).**

   You can turn the chop from side to side as you work.

3. **Cut around all sides of the bone, scraping with your knife until the bone is clean and white.**

4. **Trim the fat so that it is even, square up your corners, and feel proud.**

Figure 12-23: Frenching the bone-in rib eye.

Illustration by Wiley, Composition Services Graphics

Chuck Short Ribs (Bench)

The chuck short ribs are almost ready for the case. The section of chuck ribs you sawed off the forequarter (refer to the earlier section "Separating Out the Chuck, Arm, and Brisket from the Plate and Rib") had five bones on it; you can remove the first rib (the one that was closest to the sternum), leaving you with a four-rib slab, as shown in Figure 12-24. Simply slice under the fifth bone to remove it from the short ribs.

Figure 12-24:
Removing
the first rib.

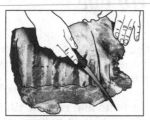

Illustration by Wiley, Composition Services Graphics

Traditionally, bone-in short ribs are *buzzed,* cross-cut on a bandsaw to the butcher's desired thickness. But because the instructions in this book use hand tools only (that is, no bandsaw), you have to either cut them in strips parallel with the bone or debone the slab completely to produce boneless chuck short ribs.

Fabricating the chuck roll

The chuck roll is a large boneless muscle system within the chuck that spans from the neck down to the fifth rib and is located under the blade bone. To extract the chuck roll, you must remove the blade bone, neck, neck bones, spine, and ribs.

To create the chuck roll, follow these steps, shown in Figure 12-25:

1. **Lay the chuck roll on the cutting surface chine down in front of you so that you can see the exposed shoulder blade (this is the place from which you previously pulled the flat iron) (Figure 12-25a).**

2. **Examine the top or outside of the shoulder blade to find a section of bone called the *scapula spine.* You'll begin your cut here.**

3. **Make a slice against the bone on the side of the muscle. Pull the meat away with your hook and scrape the bone with the back of your knife blade to release the muscle from the bone (Figure 12-25b).**

   When the muscle on top of the scapula has been cut free of the bone, you can work on removing it.

4. **Turn the chuck on its side, blade bone up. Slice against the under-side of the shoulder blade. Continue cutting in the manner instructed in Step 3, slicing and scraping as you work underneath the blade bone (Figure 12-25c).**

   In between cuts, grab the blade bone and pull it away from the muscle with your hands so that you get a better view of the under-side of the blade bone. Doing so also assists you in pulling the bone free from the chuck.

5. **Grab the shoulder blade with your free hand and pull up, slicing under the bone beneath the socket to disconnect any meat still attached to it (Figure 12-25d).**

6. **Grab the socket and yank the blade bone back to rip the scapula free.**

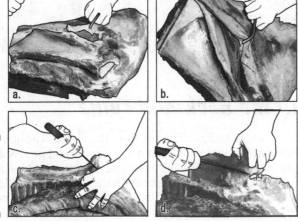

Figure 12-25:
Fabricating
the chuck
roll.

Illustration by Wiley, Composition Services Graphics

Seaming out the mock tender

The mock tender, also called the chuck tender, lies on the top inside of the chuck. This muscle is located to the side of the scapula spine you removed in the preceding section.

REMEMBER

Don't mistake this cut for being a substitute for tenderloin. The mock tender is usually braised or roasted. If you do intend to cut thin medallions and grill the mock tender, marinate it first.

To cut out the mock tender, follow these steps, shown in Figure 12-26:

1. **Locate the mock tender and use your meat hook to pull it back toward your body to find the seam (Figure 12-26a).**

2. **Make small slices with the tip of your boning knife at the seam of the muscle to begin releasing it from the chuck (Figure 12-26b).**

    Now you have a clear view of the muscle and can easily identify where the oblong-shaped mock tender begins and ends.

3. **Cut the mock tender free from the chuck (Figure 12-26c).**

4. **Remove any fat, sinew, or silver skin from the mock tender cut.**

Figure 12-26:
Cutting the
mock
tender.

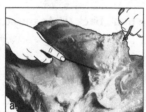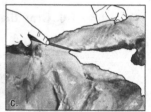

Illustration by Wiley, Composition Services Graphics

Removing the neck and spine

Time to remove the spine and neck bones from the chuck. The spine is a cumbersome, intricate bone system with many curves and grooves that make it difficult to butcher well. The task is to remove the spine and neck bones in one piece without leaving too much meat on the bone and wasting it. To accomplish this, you need to do a good amount of prep work by scoring around each vertebra and cutting as much meat free as you can. Doing so "loosens" the spine from the flesh, allowing you to cut it free when you are ready.

Expect to spend some time working around the vertebrae and down the spine, but don't feel overwhelmed. Put your deboning skills to the test with the instructions in this section.

Follow these steps, shown in Figure 12-27:

1. **Lay the soon-to-be boneless chuck roll on the cutting surface in front of you, spine up.**

2. **Pull the meat under the neck vertebrae down with your hook. Slice under the vertebrae, focusing on the neck meat before moving down to the end of the spine. Cut as far under the spine as you can (Figure 12-27a).**

   Position the bone-in chuck subprimal so that it hangs slightly off the edge of the cutting board as you work. Let gravity help you.

   Notice where the spine changes from the curved part (the neck) to the straight portion of the spine; the ribs begin here.

3. **Score around each rib, disconnecting the meat from the rib bones (Figure 12-27b).**

4. **Notice the featherbones. Pound the meat to the side of the featherbones down with your fist (Figure 12-27c).**

   The featherbones (thoracic vertebrae) extend vertically from the spine. They look like a set of thin ribs sticking up from the top of the spine.

   Pounding the meat to the side pulls the vertebrae away from the muscle so that you don't have to use your knife to work them free. (If you're frustrated from struggling with the spine, this is your chance to let loose!)

5. **Slice underneath the featherbones, moving your knife up the spine, cutting against the bone, to the cervical vertebra in the neck. Cut around the neck vertebrae (Figure 12-27d).**

6. **Pull back on the featherbones and again slice against the base of the spine, loosening the spine and neck bones even further (Figure 12-27e).**

   At this point, the prep work is done, and you're ready to remove the spine and neck bones.

7. **Lift up on the rib end of the spine and pull the chuck up off the table. While gravity pulls the meat downward, cut under the bones, vertebra by vertebra, until you've removed the spine completely (Figure 12-27f).**

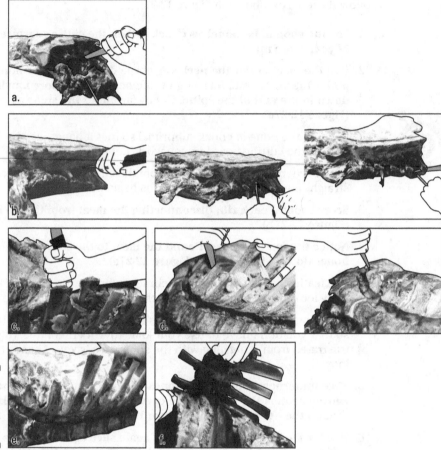

Figure 12-27:
Removing
the neck
and spine.

The last stages of the chuck

Now you have a boneless chuck roll, but you still have some neck meat
attached, which you need to remove if you want to meet industry specs.
You use a straight cut, more or less parallel to the blade end of the chuck, to
remove the neck meat.

You also have to remove the *back strap,* the thick, yellow tendon that runs
along either side of the spine. To remove the back strap, simply slice around
it, starting with one edge and remove it in one whole piece.

On the Bench: The Plate

The plate contains seven ribs and produces short ribs, boneless or bone-in, as well as the inside skirt. The meat on this primal can be ground or cubed for stew meat.

Removing the inside skirt

The inside skirt lies on top of the ribs at the bottom edge. It's covered by a layer of membrane that you have to remove, using the same technique you used to remove the membrane from the outside skirt (refer to the earlier section "Removing the Outside Skirt" for instructions).

To remove the inside skirt, follow these steps, shown in Figure 12-28:

1. **Lay the plate on the cutting surface in front of you, rib side up. Slice under the inside skirt, starting at one edge of the rib (Figure 12-28a).**

2. **Peel back the inside skirt, slicing against the base of the ribs to remove the muscle in one whole piece (Figure 12-28b).**

3. **Clean the skirt steak by removing the membrane, sinew, and if you desire, the fat.**

Figure 12-28:
Removing
the inside
skirt.

Illustration by Wiley, Composition Services Graphics

Cutting the short ribs

Short ribs are typically three to four rib sections of the plate, cut cross-wise about 1 inch thick. The plate has seven ribs, so one portion will have three ribs, and one portion will have four ribs. As I mentioned earlier, if you don't have access to a bonesaw, you may want to cut boneless short ribs after this step.

An efficient and waste-free workspace

Whether you are butchering at home or in a professional setting, the life of the animal and the value of the meat should be respected. Don't be wasteful. An experienced butcher has a plan for further processing and sets up his work station to quickly make best use of *all* of the meat as it comes off the carcass.

Cubing stew meat and setting aside trim for grind or bones for stocks happens throughout the entire butchery process. Make several piles on the table while you're working. The first is for bones, used for stock or dog bones. Another pile is for fat, which you can use for rendering or adding to grind. And then one pile is for grind. The meat in the grind pile should not be too bloody or dark. A final pile is for stew meat, larger sections of meat that come off during trim work.

To cut the short ribs, follow these steps, shown in Figure 12-29:

1. **Lay the plate on the cutting surface in front of you, ribs up.**

2. **Count four ribs and make a straight slice through the meat between the ribs (Figure 12-29a).**

3. **Stop when you reach the base of the ribs, where they connect to the breast-plate, or sternum.**

4. **Turn your knife 90 degrees and cut through the cartilage separating the ribs from the sternum (Figure 12-29b).**

5. **Cut the first section of short ribs free from the plate (Figure 12-29c).**

6. **Cut between the cartilage and the ribs, following the trajectory of the last cut to remove the second slab of short ribs (Figure 12-29d).**

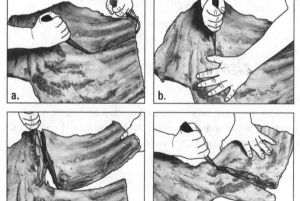

Figure 12-29:
Cutting the
short ribs.

Illustration by Wiley, Composition Services Graphics

Cleaning the breastbones

Clean the remaining portion of bones from the plate: Trim any fat or sinew from the bones, cut the pieces into smaller portions, and scrape the bones clean. Use the meat and fat for grind, and add the bones to your stock pile.

Chapter 13

Beef: The Hindquarter

In This Chapter

▶ Getting familiar with the four hindquarter primals

▶ Preserving whole muscle groups as you break the hindquarter

▶ Producing cuts ready to cook or put in the case

T he hindquarter is the posterior quarter of a carcass or half of a side of beef. It is made up of the leg, lower back (including the lumbar ribs), and the flank (belly) of the animal. In this chapter, I tell you how to break a beef hindquarter on the rail, using the same procedures and the same equipment you use to break the forequarter, explained in Chapter 12.

As I point out in other chapters, many different methods to butchering a hindquarter exist, and the method I cover here is just one of them. Bascially, you'll cut as much as possible on the rail and then move to the bench to finish up the subprimal and trim work. By following the order and techniques I give you here, you'll be able to keep whole pieces intact instead of tearing through muscles.

As you pull pieces off the hindquarter, maintain a positive work flow; refrigerate larger sections until you are ready to work on them; and separate trim, fat, and bones as you cut. You need a boning knife, a bonesaw, a butcher knife or large cimeter, and a sharpening steel. Make sure your knives are sharp and that you hone your blades throughout the butchery process to make your work easier and more efficient.

Breaking the Hindquarter: The Basics

The hindquarter has four primals: the loin, the sirloin, the round, and the flank (see Figure 13-1). In the later sections, I take you through breaking the hindquarter into primals (primary cuts), then subprimals (secondary cuts), and finally into retail cuts (the cuts you're probably most familiar with). Table 13-1 lists the cuts you'll produce in this chapter and indicates whether you make them on the rail or on the bench. (*Note:* This list is incomplete because there are just too many retail cuts to cover in one chapter.)

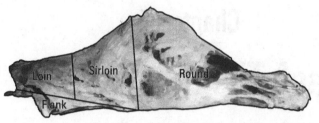

Figure 13-1:
The hindquarter primals.

Illustration by Wiley, Composition Services Graphics

Table 13-1 Primal, Subprimal, and Retail Cuts from the Hindquarter

Primals (on the Rail)	Subprimals (on the Bench)	Retail Cuts (on the Bench)
Loin	Short loin, tenderloin, strip loin or top loin	T-bone, porterhouse steak, filet, New York or strip loin, club steak (bone-in strip loin steak), hanger steak
Sirloin	Top sirloin, bottom sirloin, tri-tip tenderloin	Top sirloin steak, top sirloin roast, bottom sirloin, tri-tip steak, tri-tip roast, head filet
Round	Top round, bottom round, gooseneck	Top round roast and steak, eye of round roast and steak, knuckle, round tip roast and steak, ball-tip steak, shank, ground beef, kabobs
Flank	Flank	Flank steak, trim, bavette, inside skirt

Think of the carcass as an intricate, three-dimensional puzzle: As the pieces start to fall away, what is left changes. Each time one piece is removed, the outcome becomes more narrow and more specific to the butcher. This is why we have industry standards that help to catalog and identify cuts. And that is why knowing what you want to end up with, before you begin, is so important!

Removing the Elephant Ear (Rail)

The elephant ear is a tough cut of meat that lies on the outside layer of the hindquarter. It spans from near the hip to the underside of the belly and is also known as *rose meat*. Remove this muscle first. After doing a little trim work, you can cut it into strips for grind, stuff and braise it, fry and mince it for tacos, or set it aside for trim.

The elephant ear is often used in Latin cooking, and in Argentina, it's an ingredient in a well-loved dish called *matambre*.

To remove and trim the elephant ear, follow these steps (see Figure 13-2):

1. **Find the elephant ear and use the tip of your boning knife to score around the muscle on either side (Figure 13-2a).**

   The elephant ear is the outside of the hindquarter, near the flank. It's a dark triangular-shaped muscle that lies on the surface of the carcass.

2. **Grab the top corner of the elephant ear with your boning hook. Pull down on the muscle and cut at the seam. Continue pulling down on the muscle and slicing at the seam until the elephant ear is removed (Figure 13-2b through d).**

   Move your boning hook in as you cut, inserting your hook closer to the seam as you move down the muscle. Doing so helps you pull the elephant ear free in order to remove the muscle in one whole piece.

   The elephant ear most likely dried on the outside as the meat aged or was exposed to air during transport. On the inside of the muscle is a bit of fat that you need to trim off. ***Note:*** You do this bit of work on the bench.

3. **Set the elephant ear inside up on the cutting surface in front of you and trim the fat from the surface of the meat (Figure 13-2e).**

4. **Score the top of the elephant ear meat in a cross-hatch pattern (Figure 13-2f).**

   This move lets you release the outer seam of the cut so that you can remove it easily without cutting too far inside or outside the muscle.

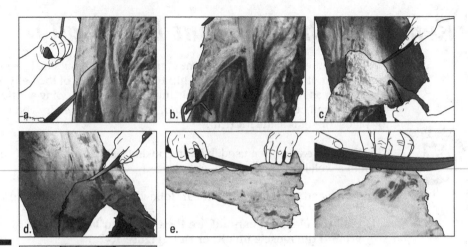

Figure 13-2:
Removing
and trim-
ming the
elephant
ear.

Illustration by Wiley, Composition Services Graphics

Pulling the Cod Fat (Rail)

The cod fat is the soft fat that surrounds the muscles in the hindquarter. Typically you render it for lard. To trim the fat, follow these steps (see Figure 13-3):

1. **Grab the fat with your boning hook directly above the area where you just removed the elephant ear (Figure 13-3a).**

2. **Slice into the fat and begin pulling it down with your boning hook (Figure 13-3b).**

   Separate the fat at the seam, making small slices with your boning knife, and pull it off in one large piece.

   You can easily identify the seam separating the fat from the muscle that lies underneath it. As you cut, you'll see the muscle. Don't cut into it; just slice on top of it, at the seam.

Figure 13-3:
Pulling the
cod fat.

Illustration by Wiley, Composition Services Graphics

Dealing with the Flank

Flank steak is a well-known grilling cut that comes from the flank primal. Flank steak is best grilled medium rare and cut against the grain (90 degree angle) when served. This steak takes well to marinades and is great sliced thin for fajitas or stuffing and rolling. In the following sections, I explain how to remove the flank and then how to prepare it for the case or cooking.

Removing the flank (rail)

The flank is a flat, boneless section of muscle on the bottom side of the hindquarter opposite the spine. This piece of meat is an easy one to remove. Just follow these steps, shown in Figure 13-4:

1. **Locate the flank (Figure 13-4a).**

   The flank is like a thick flap of meat. If you look at the hindquarter, you can see the dark muscle (the flank); beside it, you can see a white space (the seam). You simply cut in-between this area.

2. **Grab the outside top edge of the flank with your boning knife and slice through the meat in between the muscle and the seam to remove it in one whole piece (Figure 13-4b).**

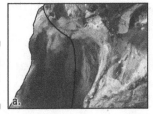

Figure 13-4:
Removing
the flank.

Illustration by Wiley, Composition Services Graphics

Freeing the flank steak (bench)

The flank is surrounded by fat, sinew, and serous membrane. You need to do a bit of trim work before this popular cut is ready for the case. When you clean a flank, you first need to cut the fat and sinew from one side ("unlocking it") before you can pull the muscle free with your hands and boning knife. You perform this work on the bench.

Follow these steps to get the flank steak ready for the case or for cooking (see Figure 13-5):

1. **Lay the flank down on the cutting board in front of you; cut about a 1/2 inch off the thicker side of the muscle (Figure 13-5a).**

2. **Grab the membrane from the top edge of the flank and rip it off.**

   Doing so removes the dried membrane from the flank in one whole piece. Now you can clearly see the flank steak.

3. **Trim some fat from the top edge of the flank at the seam (Figure 13-5b).**

4. **Score around the top corners of the flank (Figure 13-5c).**

   Scoring around the top corners lets you to grab the steak with your hands in the next step and separate it from the serous membrane underneath.

5. **Lift up on the top edge of the flank steak and pull it back and away from the membrane with your hands (Figure 13-5d).**

   To make this task easier, place your free hand on top of the thick, serous membrane as you pull.

6. **When you can no longer pull the muscle free with your hands, use your boning knife to slice against the seam (Figure 13-5e) and then cut the remaining portion of the flank free from the outside membrane.**

7. **Trim any remaining meat from the membrane and set the trimmed meat aside for grind.**

8. **Trim the fat from the top and sides of the flank and square off the bottom edge (Figure 13-5f and g).**

   Now your steak is ready to prepare and serve or to put in the meat case.

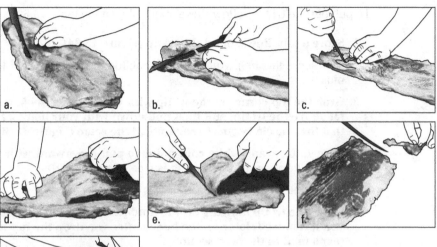

Figure 13-5:
Freeing and
cleaning the
flank steak.

Illustration by Wiley, Composition Services Graphics

Pulling the Tri-Tip (Rail)

The tri-tip is a boneless, single muscle located at the bottom of the sirloin butt opposite the spine. This small, triangular-shaped muscle usually weighs around 1½ to 2½ pounds. The tri-tip can be used whole, cut into steaks, or cubed for kabobs. It's a California barbecue favorite and takes well to marinades and dry rubs.

Although the traditional way to deal with the tri-tip is to cut through it to separate the sirloin and the round, in these steps, I have you pull the tri-tip from the hindquarter on the rail. Here's why:

✔ This method lets you to pull the tri-tip in one whole muscle group instead of cutting through it.

✔ It gives you a clear view of the knuckle or sirloin tip so that you can cut around it, leaving it intact when you break the hindquarter in the later sections.

To pull out the tri-tip, follow these steps, shown in Figure 13-6:

1. **Score a thin line around the tri-tip (Figure 13-6a).**

   This score line help you feel more confident when you start to pull the muscle.

2. **Grab the top corner of the tri-tip with your boning hook, cut into the fat above the tri-tip, and then pull down with your boning hook so that the muscle begins to separate at the seam (Figure 13-6b and c).**

3. **Slice at the seam, while continuing to pull downward with your boning hook, until the muscle is free (Figure 13-6d).**

   Move your boning hook in closer as you work.

4. **Leave the tri-tip hanging off the hindquarter for now (Figure 13-6e); you'll remove it later on the bench after you drop the full loin (explained in the next section).**

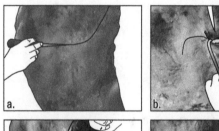 

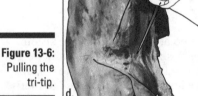 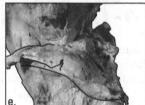

Figure 13-6:
Pulling the
tri-tip.

Illustration by Wiley, Composition Services Graphics

Removing the Full Loin (Rail)

The full loin is the primal portion that includes the sirloin and the loin, or short loin. It spans from the cut end of the ribs to the lower, middle end of the animal's back. In this section, I tell you how to separate the full loin from the round.

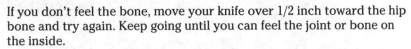

Follow these steps, shown in Figure 13-7:

1. **Locate the hip joint and insert your boning knife into the joint at the bottom of the round, about 1 inch down and a few inches in from the hip bone (Figure 13-7a).**

   The hip joint is approximately 1 inch in from the tip of the hip bone, which you can easily see from the outside surface of the hindquarter.

2. **With your knife still inserted, quickly tap it in and out to make sure you "feel" bone on the tip of the blade.**

   If you don't feel the bone, move your knife over 1/2 inch toward the hip bone and try again. Keep going until you can feel the joint or bone on the inside.

   After you find the joint, you can do the cut work necessary to prepare for sawing through the bone and dropping the full loin from the hindquarter.

3. **Switch to a large cimeter or butcher knife and insert the tip into the same spot on the joint.**

   "Tap" again to make sure you're in the right spot.

4. **Make a straight cut from the hip joint through the tail of the tenderloin (this is the small section of tenderloin muscle) and then stop (Figure 13-7b).**

   From this point, your knife cut changes direction.

5. **Cut down under the knuckle and then straight (parallel to the cut end of the loin) across the hindquarter and ending at the spine (Figure 13-7c).**

   During this cut, keep the tip of your large butcher knife tapping against the hip joint as you make small, ever-so-slight sawing motions with your blade.

6. **Using your bonesaw, saw through the bone and then stop! Don't saw through the meat (Figure 13-7d).**

   At this stage, the full loin will still be hanging from the hindquarter (Figure 13-7e). You want to position yourself properly before you cut it free.

7. **Move yourself into position so that you can catch the loin after you make the last cut to free it from the round (Figure 13-7f). Cut through the muscle attaching the loin and sirloin to the round.**

   When it falls, the loin will weigh around 73 pounds (approximately 17 percent of the animal's hanging weight), so make sure to get a good grip on it before you cut it free. Stand to the outside of the loin, use your free arm to hold the whole loin with your arm on the side of your body. You can grab under the kidneys with your free hand.

8. **Set the full loin aside under refrigeration until you're ready to process it further.**

   For instructions on processing the loin, head to the later section "Cutting the Full Loin Down."

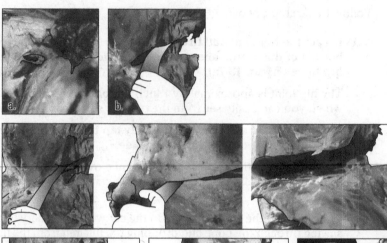
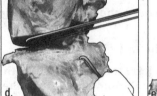

Figure 13-7:
Separating
the loin and
sirloin from
the round.

Removing and Portioning the Round (Rail)

Within the round primal, you have these subprimals: the gooseneck (bottom round), the knuckle (round tip), the full cut round, and the top round. From these subprimals, you can produce many different lean steaks and roasts that are reasonably tender and often underutilized. A slow-cooked, medium-rare roast from the round makes a delicious roast beef, and tougher cuts can be sliced thin into steaks and marinated to perk up on the grill or in a quick sauté. In this portion of the cow, you'll also find plenty of lean stew meat, grind, and good marrow and stock bones. In the following sections, I explain how to section the round from the carcass.

Removing the knuckle from the round

The knuckle is located on the front of the round, spanning from the knee to the hip. From the knuckle, you can produce cross-wise cut steaks (ball tip steaks and round tip steaks), tie roasts, or seam the muscles out even further to produce smaller roasts and steaks.

You separate the knuckle from the top (inside) round and the bottom (goose-neck) round by cutting along the natural seams.

Part of the knuckle is already exposed where you removed the tri-tip (refer to the earlier section "Pulling the Tri-Tip"). This earlier work can help you identify the seams on either side of the knuckle. You'll follow these seams to separate the top and bottom round.

To remove the knuckle, follow these steps, shown in Figure 13-8:

1. **Find the natural seam on either side of the knuckle and slice the seams open (Figure 13-8a).**

2. **Grab the top of the knuckle with your boning hook and make a slice on top of the knee joint where the femur and tibia connect (Figure 13-8b).**

3. **Pull down on the knuckle with your boning hook — put some muscle into it! — and use your boning knife to assist the separation, slicing down the seams on both sides (Figure 13-8c).**

   As you pull, the muscle will begin to pull apart at the seams.

4. **Cut the knuckle off the round (Figure 13-8d and e) and set it aside under refrigeration until you are ready to process it into retail cuts.**

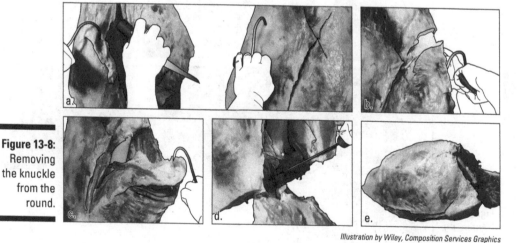

Figure 13-8:
Removing the knuckle from the round.

Illustration by Wiley, Composition Services Graphics

Cutting the top sirloin free from the round

The top round is the muscle from the inside of the leg, or inner thigh region. You can cut the top round into thick steaks, tie it into roasts, cube it for stew meat, grind it, or make it into kabobs and marinate it for the grill.

To cut the top sirloin free from the round, follow these steps, shown in Figure 13-9:

1. **Use your boning hook to grab the top corner of the top sirloin; then pull down with your hook and make a small slice into the meat (Figure 13-9a).**

2. **Cut the top edge of the top sirloin free by cutting at the seam on the right and then on the left (Figure 13-9b).**

   As you pull with the hook and cut along the seams, the muscle will start to separate.

3. **Now that your cut is started, pull down with your boning hook and cut the top sirloin off the round by separating it at the natural seams (Figure 13-9c).**

4. **Cut the top sirloin free from the round (Figure 13-9d).**

   You can now set the top round aside, under refrigeration until you're ready to cut it into retail cuts or prepare and serve it.

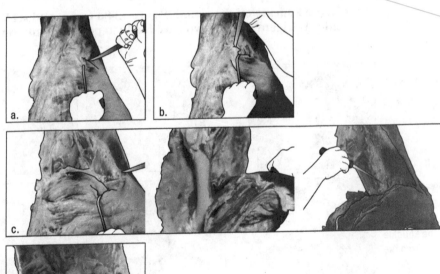

Figure 13-9: Cutting the top sirloin free from the round.

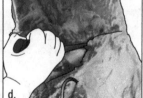

Illustration by Wiley, Composition Services Graphics

Removing the gooseneck (bottom round)

The gooseneck, or bottom round, is the outer muscle on the rear leg. The cuts on this subprimal are lean but not very tender. Still, this part yields many roasts and steaks, like the eye of round, rump roast, round steaks, and heel.

You remove the gooseneck from the round at the natural seams. Follow these steps (see Figure 13-10):

1. **Grab the top of the gooseneck with your boning hook and make a slice at the top of the muscle (Figure 13-10a).**

2. **Cut around the side of the gooseneck to release the seam (Figure 13-10b).**

3. **Pull down on the muscle with your boning hook and continue to slice at the seam with your boning knife (Figure 13-10c).**

4. **When the gooseneck is pulled down and off the bone, cut it free (Figure 13-10d).**

   Set the gooseneck aside under refrigeration until you are ready to process it into retail cuts. From this subprimal, you can produce bottom round roasts, steaks, or rump roast, along with stew meat and grind.

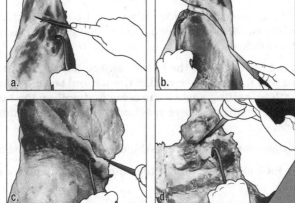

Figure 13-10:
Removing the goose-neck, or bottom round, from the round.

Illustration by Wiley, Composition Services Graphics

Cutting the Full Loin Down (Bench)

Because you removed the flank steak before you dropped the full loin, a portion of the flank primal still remains on the loin. The flank muscle system has several muscle layers, including the inside skirt, the bottom sirloin flap (also called the bavette), and some fattier flap meat that is great for grind.

To remove the section of flank that spans from the bottom of the sirloin (the tri-tip end) to the last rib (number 13) on the short loin, you peel back the tip of the bottom sirloin flap, cut the remaining flank from the loin, and then seam out the inside skirt and the bottom sirloin flap. The rest you reserve for grind. The following sections provide the details.

Removing the flank from the full loin

In this section, I explain how to pull back the tip of the bottom sirloin flap and cut the remaining flank off. By removing the muscle groups intact, you can seam them out into whole pieces.

Follow these steps, shown in Figure 13-11:

1. **Lay the loin inside up on the cutting board in front of you and find the top corner of the bottom sirloin flap (Figure 13-11a).**

    At the cut end of the sirloin, notice where the tip of the bottom sirloin flap lays on top of the tri-tip, which you left "hanging" off the sirloin end (refer to the earlier section "Pulling the tri-tip").

2. **Pull up on the top corner of the bottom sirloin flap with your free hand and, as you pull it back, slice against the natural seam (Figures 13-11b and c). Continue cutting against the seam and pulling the bottom sirloin flap back until you have reached the point where the tri-tip muscle ends.**

3. **When you have cut past the tri-tip (Figure 13-11d), stop seaming.**

4. **Now take a look at the cut end of the short loin and make a notch on the 13th rib.**

    This is where you want to separate the flap meat from the loin. The cut you make here determines how much "tail" you leave on your short loin.

5. **Hold your knife in a dagger grip and, starting at the point where the tri-tip ends, make a vertical cut parallel to the spine, ending at the notch you made in Step 4 (Figure 13-11e).**

6. **Saw through the rib bone to remove the flap meat in one clean piece (Figure 13-11f).**

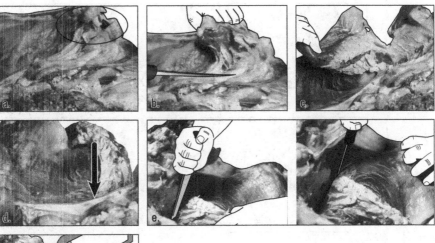

Figure 13-11:
Cutting the
flank from
the loin.

Illustration by Wiley, Composition Services Graphics

Seaming out the inside skirt

Now you are ready to seam apart the flank muscles. This crafty cut work isolates the remaining portion of the inside skirt and the bottom sirloin flap. You'll clean the rest for grind and also remove the 13th rib. When you're done, you'll have the inside skirt. Although this cut is not as tender as the outside skirt, you can marinate it or slice it thin against the grain for fajitas.

Follow these steps (see Figure 13-12) to free the inside skirt:

1. **Lay the flank section inside up on the cutting surface in front of you. Examine the top to find the small portion of inside skirt and score around it (Figure 13-12a).**

2. **Remove the membrane from the top of the inside skirt (Figure 13-12b).**

3. **Peel up one corner of the inside skirt and pull it away from the flap with your fingers; you can use the tip of your boning knife to help release it (Figure 13-12c).**

4. **Pull the skirt back as you slice it free at the seam; then cut the inside skirt from the flap (Figure 13-12d).**

5. **Clean up the skirt by trimming the edges and sides of fat or sinew (Figure 13-12e).**

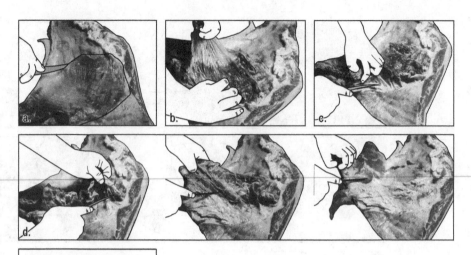

Illustration by Wiley, Composition Services Graphics

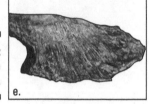

Figure 13-12:
Freeing the
inside skirt.

Seaming out the bottom sirloin flap (bavette steak)

The bottom sirloin flap, or *bavette steak,* is not as well known as other cuts, but it's rich in flavor and has great texture. In fact, it's one of my favorite cuts. The bavette steak is similar in characteristic to flank or skirt steak and is great when quickly seared or grilled and cut against the grain.

To free the bottom sirloin flap, follow these steps, shown in Figure 13-13:

1. **Trim some of the fat and silver skin on top of the bottom sirloin flap (Figure 13-13a).**

2. **Score alongside the 13th rib and continue the trajectory of your scoring cut to curve under the bottom edge of the bottom sirloin flap (Figure 13-13b).**

3. **Grab the tail end of the bottom sirloin flap and peel it back. Use your hand and your knife together to release the muscle, pulling and cutting at the seam as you work until it is free (Figure 13-13c).**

4. **Remove any silver skin and clean up the edges on both sides of the muscle (Figure 13-13d).**

   After you trim any remaining flap meat and set it aside for grind, the bottom sirloin flap is now ready to use. Yum!

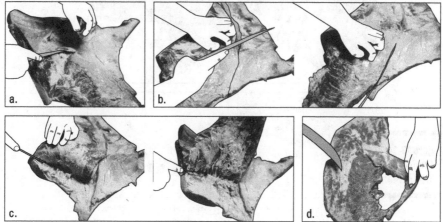

Figure 13-13: Removing the bottom sirloin flap.

a. b. c. d.

Illustration by Wiley, Composition Services Graphics

Cutting the Tri-Tip Free (Bench)

The tri-tip is, at this point, pretty much ready to cut free from the sirloin. After you do so, you simply need to trim some fat off of it to make it ready for the case or cooking.

Follow these steps, shown in Figure 13-13, to cut the tri-tip free:

1. **Cut the tri-tip from the sirloin at the seam (Figure 13-14a).**

   Be careful not to cut too far into the sirloin. The head filet is located next to the tri-tip on the sirloin. Try not to cut into that muscle.

   Position the sirloin end of the loin so that the tri-tip hangs off the table while you cut. Use your boning hook to maintain control of the muscle while you make small slices with the tip of your boning knife at the seam. Gravity will help pull the tri-tip down, allowing you to clearly see where the whole muscle separates from the sirloin.

2. **Lay the tri-tip down on the cutting surface in front of you and trim any fat, sinew, or silver skin from the bottom of the muscle (Figure 13-14b).**

3. **Trim and square up the edges of the tri-tip (Figure 13-14c).**

TIP

Whether you leave some fat on the top of the tri-tip is up to you. Industry cut standards exclude the fat, but I prefer it, so I make sure the bottom looks nice and clean.

Figure 13-14:
Cutting the
tri-tip free.

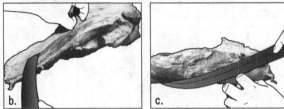

Illustration by Wiley, Composition Services Graphics

Separating the Short Loin from the Sirloin (Bench)

Now that you've removed the bottom sirloin flap meat and the tri-tip, you can separate the short loin from the sirloin. The short loin gives you T-bone, porterhouse, and club steaks. You can also debone the sirloin and seam apart the muscle groups to produce a variety of steaks and roasts.

To remove the short loin from the sirloin, follow these steps, shown in Figure 13-15:

1. **Find your cut line in the center of the lowest lumbar vertebra just above the curved joint (Figure 13-15a).**

   If you look at the spine, you can clearly see where the spine curves and changes from the lumbar to the sacral vertebra. Identify about 1 inch just above the first curved vertebra, in the center of the lowest lumbar vertebra, as your cut line.

2. **Using your large butcher knife or cimeter, cut down through the meat in a straight line from your cutting point (Figure 13-15b).**

3. **Following your cut line, saw through the bone and any remaining meat (Figure 13-15c).**

Figure 13-15:
Removing
the short
loin fro the
sirloin.

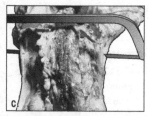

Illustration by Wiley, Composition Services Graphics

Taking Care of the Top Sirloin (Bench)

From the sirloin primal, you get top sirloin, bottom sirloin, ball tip, tri-tip, and bottom sirloin flap. You have already removed the bottom sirloin flap and the tri-tip, so all you have left to do is to remove the head filet (or butt) of the tenderloin and debone the top sirloin, as I explain in the following sections.

Removing the head filet

To remove the head filet, follow these steps, shown in Figure 13-16:

1. **Score against the top of the bone with your knife. Pull up on the top side of the muscle (head filet) and scrape your knife against the bone (Figure 13-16a).**

2. **Lift up the head filet from the opposite side at the lower seam and score against the bone underneath the muscle (Figure 13-16b).**

   These scoring cuts prepare you for removing the head filet in the next step, leaving behind a clean, white bone.

3. **Pull up on the head filet from the top and cut underneath it, scraping against the bone (Figure 13-16c).**

4. **Peel back the head filet, scraping the bone and cutting as you move down the length of the femur (Figure 13-16d). Cut the muscle free.**

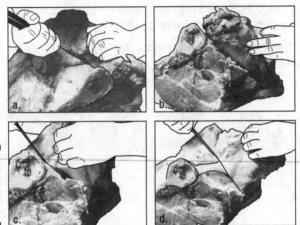

Illustration by Wiley, Composition Services Graphics

Figure 13-16:
Removing the head filet.

Deboning the top sirloin

To debone the top sirloin, follow these steps, shown in Figure 13-17:

1. **Pull down on the top sirloin with your boning hook near the bottom of the hip bone and scrape against the bone with your knife (Figure 13-17a).**

2. **Scrape around both sides of the hip bone until the surrounding meat is loosened from the bone (Figure 13-17b).**

3. **Turn the bone-in top sirloin spine up on the cutting surface in front of you and cut against bones along the spine (Figure 13-17c).**

4. **Pull the hip bone away from the muscle, cutting it free with your boning knife as you work (Figure 13-17d).**

With the bone removed, the top sirloin is ready to set aside, under refrigeration, until you are ready to process it into retail cuts, like sirloin steaks, butt steaks, or a tied sirloin roast (after you clean off the fat and silver skin).

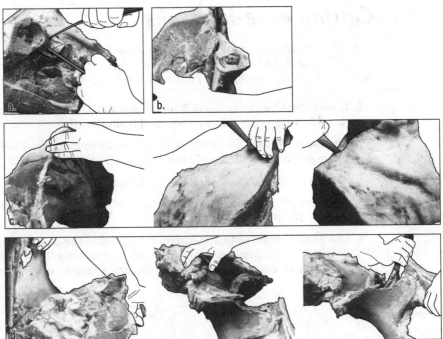

Figure 13-17:
Deboning
the top
sirloin.

Illustration by Wiley, Composition Services Graphics

Cutting Steaks from the Short Loin (Bench)

The short loin gives you several options for bone-in or boneless steaks. On one side of the lumbar vertebra, you have the tenderloin, and on the other side, you have the top loin, or strip loin, which is well known for tasty boneless New York steaks. In this section, I explain how to cut bone-in steaks by hand with a bonesaw.

As the tenderloin moves up the short loin, the names of the bone-in steaks change:

- ✔ **Porterhouse steaks** are cut from the rear end of the short loin, where the portion of tenderloin is largest.

- ✔ **T-bone steaks** come from the area where the tenderloin becomes smaller.

- ✔ **Club steaks,** which are essentially bone-in New Yorks steaks, come from the area where the tenderloin ends.

Cutting bone-in steaks

Start with the rib end of the short loin and cut a couple of bone-in club steaks. To cut the club steak from the short loin, you need a bonesaw and a large butcher knife or cimeter. Follow these steps, shown in Figure 13-18:

1. **If your beef has been aged, trim off any dark crust or dried meat from the side where the loin was severed from the rib.**

2. **Choose how thick you want your first bone-in club steak to be and cut through the meat alongside the 13th rib (Figure 13-18a).**

   The first steak you cut will have the 13th rib on it.

3. **Saw through the bone to remove the club steak from the short loin (Figure 13-18b).**

4. **Continue cutting in this manner, determining the desired thickness of the steak, cutting through the meat, and sawing through the bone, until you've cut the whole short loin into a variety of bone-in chops.**

Figure 13-18:
Cutting steaks from the short loin.

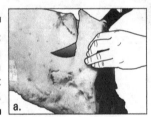

Illustration by Wiley, Composition Services Graphics

Frenching the bone-in steaks

For a more dramatic presentation, impress your friends and neighbors with your crafty cuts by frenching the bone-in chops. Follow these steps, shown in Figure 13-19:

1. **Hold the bone-in steak with your free hand and rest the rib bone on the table in front of you, meat side up.**

2. **Cut about halfway down the bone and into the meat; then turn your knife blade away from you and scrape down the bone, removing some of the rib meat (Figure 13-19a).**

   Don't cut too far down; leave a little of that tasty fat and meat on the bones!

   You can turn the chop from side to side as you work.

3. **Cut around all sides of the bone, scraping with your knife until the bone is clean and white (Figure 13-19b).**

4. **Trim a thin layer of fat from the outside of the steak (Figure 13-19c).**

Figure 13-19:
Frenching
a bone-in
steak.

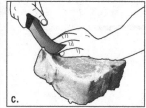

Illustration by Wiley, Composition Services Graphics

Producing Osso Bucco from the Hindshank (Bench)

At this point, you have a variety of cuts you've made on the bench from the loin and round portions, but still hanging on the rail is the hindshank with the femur bone still attached. The last task you need to do is to cut the femur bone off and then move the rest of the hindshank to the bench, where you'll produce crosscut shanks, or osso bucco. At that point, you can also debone the shank and tie a roast for braising or slow roasting under moist heat.

Cutting osso bucco can be a little tricky with a bonesaw. The trick is to make sure that the portions are fairly thick so you don't have to struggle with the saw too much and don't tear up the meat in the process.

To remove the femur and cut osso bucco, follow these steps, shown in Figure 13-20:

1. **Remove the femur by making a cut between the joint, as shown in Figure 13-20a.**

   You make this cut on the rail.

2. **Place the hindshank on the bench and then slice through the shank meat approximately 1–2 inches thick, starting from the knee side (see Figure 13-20b).**

3. **Saw through the bone with your bonesaw (Figure 13-20c).**

   Use your free hand to steady the meat so that it doesn't roll while you saw.

   When you are finished sawing, scrape or wipe the meat near the sawed bones. The sawing motion creates friction and forms a substance of emulsified fat, blood, and bone fragments on the meat that decreases shelf life.

Figure 13-20:
Cutting osso
bucco.

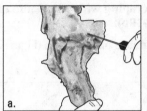
a.

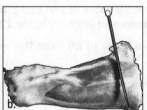
b.

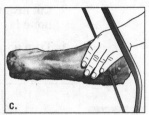
c.

Part V

Sausage-Making and Using the Whole Animal

The 5th Wave   By Rich Tennant

"Alright, this should make everyone feel a little better. It's a bowl of my own, homemade chicken farmer soup. Sip it down carefully and watch for bones."

In this part . . .

In this part, you discover what you can do with all the delicious trim and fats that are left over from the butchery process. In this part, your creativity can truly shine. After you get a handle on further processing basics, like sausage-making, curing, and smoking, you can use your knowledge to create your own special dishes to share with friends and family alike.

Chapter 14

Setting Yourself Up for Sausage

In This Chapter
▶ Finding the necessary equipment
▶ Discovering common sausage seasonings

After the butchering is done, you'll find yourself with a healthy amount of trim and cuts that are well suited for sausage-making. You can make large batches of sausage and give them to friends as gifts or freeze to enjoy later. Making sausages, pâtés, and terrines from the "leftovers" is a great way to use the entire animal and create a variety of delicious dishes and products.

In this chapter, I introduce you to the basics of sausage-making. Here you discover the kind of equipment and ingredients you need and the kinds of sausages you can make. For information on the process, head to Chapter 15. Chapter 16 has several astounding sausage recipes you can try out yourself.

Gathering the Right Equipment

Sausage is actually a pretty simple product to make. Making mistakes and figuring out how to produce a well-balanced sausage is all a part of the process and how you learn. If you plan to make sausage at home as a hobby, using the sausage stuffer attachment on your mixer may be all you need for small-batch experimenting. But in most cases, you'll quickly outgrow that equipment.

When you're ready to move up, you'll want the right equipment. Of course, the "right" equipment for you depends on your needs and budget. Many different kinds of sausage-making equipment are available. In this section, I tell you what you need to know. Armed with this information, you can determine what options may work for you, based on your experience and long-term goals.

Cost of equipment can vary greatly. As you decide how much to spend, you want to take into account the quality of the equipment, how much effort and time you plan to put into the task, and how much sausage you plan to make. If you're a home butcher, your needs will be different from those of a professional butcher, and the "best" price is not necessarily the lowest price.

When more is better: Advice for pros

Professional butchers need more than the basic equipment. Says Mark M. DeNittis, of Rocky Mountain Institute of Meat:

"The costs of business to produce and manufacture items are not cheap. Having the right equipment not only goes a long way, but it also provides savings in efficiency and employee productivity.

"Having several pieces of equipment versus one that can accomplish the same end result without sacrifice to quality and integrity of the product is of significant importance. Being able to achieve the same results with a single piece of equipment not only saves steps and time in production but also in clean up and sanitizing.

"Think of time and work related to sausage-making: the production and process time of making the sausage itself, as well as the time it takes to break down parts and pieces of the equipment and then the time it takes to clean, sanitize, and then reassemble them. If you look at these facets alone and the time involved in one shift, you may be surprised. Now multiply that by how many shifts in a week and then multiply that by 52 (the number of weeks in a year). This exercise gives you insight into the monetary value of equipment selections."

Thinking about your sausage-making needs

The kind and volume of sausage you plan to make influences the type of sausage-making equipment you need and the ultimate cost.

- ✔ **Kind of sausage (ground or link sausage):** You need a grinder and a sausage stuffer to make link sausage. You can use a mixer with a stuffing horn attachment as well. You need only a grinder if you want to produce fresh ground sausage.

- ✔ **Amount of sausage:** Depending on the size of the batch you are making, your equipment needs will change. Large grinders, sausage stuffers, and mixing bowls make ease of production, especially if you are making batches larger than 5 pounds.

Choosing a grinder

Grinders come in a variety of styles and prices (see Figure 14-1). For under $150, you can find grinders that are very popular among home enthusiasts or shops that produce less than 25 pounds of sausage a day. Here are a couple of options:

- **Hand-crank grinders:** Hand-crank grinders are an inexpensive option, but not necessarily the most efficient one. I have never had a lot of success with hand-crank grinders; they can bind up with tissue quickly and require a lot of elbow grease to grind large amounts of meat. But if you want to experiment without making a big investment, you could consider trying one out for a short period of time.

- **Electric table-top grinders:** These grinders are a bit more expensive but offer better quality. Ideally you want to purchase one with metal gears and bushings (cheap grinders have cheap parts). They're available in a variety of horsepowers and grinder head sizes. Get information from the manufacturer about what the machine can handle and choose one based on your needs (low horsepower models, for example, can bind up if you overwork the motor).

Of course, if you plan on making quite a bit more sausage than 25 pounds a day (you're a restaurant chef or a hunter, for example), you need a grinder that can handle the extra workload. Here are your options:

- **Low- to medium-volume grinders:** These can produce 100 or more pounds of sausage a day and cost $300 and upward. They're popular with restaurant chefs and hunters.

- **High-volume grinders:** These pricey models (costing over $1,500) enable you to produce over 500 or more pounds of sausage a day. Super high-volume grinders can produce more than 1,000 pounds of sausage a day, and their price ($3,000 or more) reflects the added capacity.

Regardless of what kind of grinder you buy, you want one that comes with plates. Most consumer grinders come with a 3/8- and a 1/4-inch plate. If you're lucky, you'll also get a fine-grind plate. If not, don't worry about it; the first two sizes are great for a lot of sausages.

Looking at mixers

For the home sausage-maker, you can mix your seasonings and other stuff in a kitchen bowl, using your hands. But if you're serious about sausage-making or plan to make a lot of it, you may also want to invest in a mixer.

Mixers come in a variety of sizes (small, medium, and large) and in different styles. You can get a hand-crank mixer (like the one shown in Figure 14-2) or one that's designed as an attachment for your grinder.

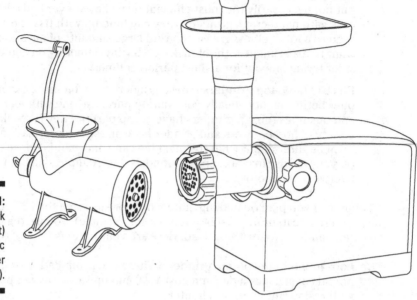

Figure 14-1:
Hand-crank
grinder (left)
and electric
grinder
(right).

Illustration by Wiley, Composition Services Graphics

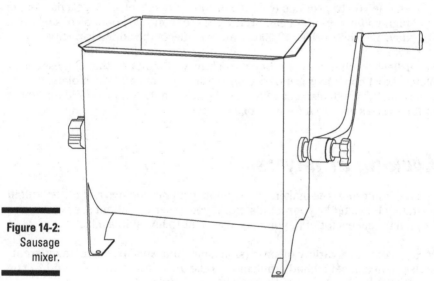

Figure 14-2:
Sausage
mixer.

Illustration by Wiley, Composition Services Graphics

Have stuffer, will sausage

Sausage patties will take you only so far. If you want to create links, brats, or other tubular sausages, you need a stuffer. Stuffers, like the one shown in Figure 14-3, enable you to put sausage in a casing.

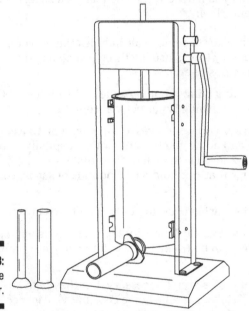

Figure 14-3:
Sausage
stuffer.

Illustration by Wiley, Composition Services Graphics

Just as grinders come in low- to high-volume models, stuffers do, too. If you're an at-home sausage-making enthusiast, you probably don't need a stuffer that can accommodate more than 5–25 pound batches of sausage. Here are some of your choices:

- **Vertical stuffer:** With these stuffers, the mixture goes in the top and comes out in casings on the bottom. You can find electric or hand crank models with capacities that range from 5–15 pounds. I recommend this type of stuffer for the home sausage-maker.

- **Counter-mounted hand stuffer:** This kind of stuffer requires some practice, and results can be less than idea.

  If you use one of these stuffers and find that you get leaks around the stuffing piston, try this trick: Place the meat paste in a sheet of plastic wrap, cut off one end to let the meat out, and squeeze the paste into the machine.

- **Hand-held stuffer:** This kind may be worth a try. I've had some that worked great, and some that didn't. They're really affordable and are available at outdoor outfitter stores. (They're great when you want to make 5-pound or less "test" batches of sausage.)

Of course, if your sausage-making endeavors are on a larger scale, make sure that the stuffer you get can meet your demands.

Other essentials

Here are a few other items you'll need when you make sausage, some of which you can see in Figure 14-4:

- ✔ **Pricker:** You use a pricker to remove air bubbles that develop as you link. This very necessary tool is usually three pronged and may have a simple knife on the opposite end.

- ✔ **Sharp knife for working trim:** A boning knife is the preferred knife for this task, but a good chef knife is perfectly acceptable.

- ✔ **Meat lug or bowl/tray:** You want a vessel large enough to easily hold the meat and spice and allow you to mix it without spilling. If you're making sausage at home, a clean dish drainer tub works great, but commercial meat lugs that can hold 45–50 pounds of sausage are also available.

- ✔ **Twine:** Quality kitchen twine is useful for certain types of sausage.

- ✔ **Stainless-steel S hook:** Even at home, drying sausages is an important step. Though a fridge can be tricky to make into a drying chamber, these hooks might help.

- ✔ **Digital scale:** Digital scales aren't necessary, but they help ensure accurate measuring, which is nice for the home cook but vital in a commercial setting like a restaurant or meat shop. If you opt for a digital scale, it needs to go to only a tenth of a gram (0.1), which is all the precision you need.

A word about casings

When you make link sausages, you'll need casings into which you'll stuff the sausage mixture. I explain how to use casings in the next chapter; here I tell you what to look for when you buy casings.

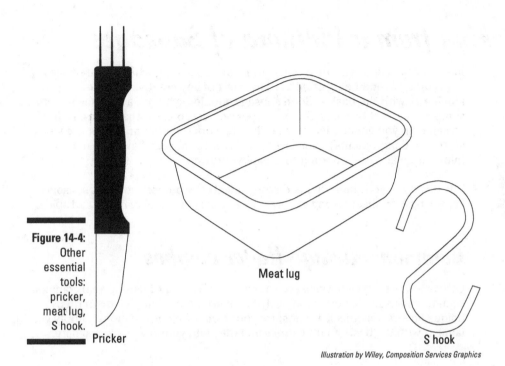

Figure 14-4:
Other
essential
tools:
pricker,
meat lug,
S hook.

Meat lug

Pricker

S hook

Illustration by Wiley, Composition Services Graphics

When making sausages, you can choose from either artificial or fresh casings, but the kind of casing you use depends on what you are making. Artificial casings are typically used for pre-cooked products like hotdogs, summer sausage, or snack sticks (these are also a good choice for kosher products), but they are not ideal for fresh sausage. Think of a casing as a barrier or protective outer layer for your sausage; the thicker the casing, the longer the sausage can dry. For example, beef middles are so thick that they're inedible, but they are the preferred casing for salumists, who intend to age their salumi for months. Fresh sausages require less protection because they are consumed quickly; fresh casings like hog and lamb are thinner and edible and ideal for these types of fresh sausages.

Make sure the casing you buy has these certain properties:

✔ It must adhere to the meat mix.

✔ It must shrink with the meat mixture during processing.

✔ It must be permeable to moisture and smoke.

Picking from a Plethora of Sausages

In sausage-making, the key to success is understanding what you are trying to accomplish and being creative. The basis of any creativity, however, is familiarity with the basics. Good sausage should be juicy, savory, meaty, and sometimes sweet or spicy. Study recipes and get to know the basics. Doing so can help you get an idea of what flavors work together and what the key spices (and ingredients) comprise your favorite sausage flavors. Use that information and build upon it to create your own.

The quality of your ingredients directly affects the outcome of your sausage. For the best, freshest-tasting sausage, always use good quality meats and spices.

Common sausage flavor combos

Quite a few basic types of sausage already exist. Table 14-1 lists several common signature sausage flavors, along with the ingredients that give them their unique flavors. Use this list to inspire your own delicious creations. Simply get familiar with the basic combinations and then add your own personal flare.

Table 14-1	Common Sausage Flavors
Signature Flavor	_Ingredients_
Andouille	Onion powder, pepper, thyme, allspice, nutmeg, cayenne pepper, sugar, paprika, ground bay leaf, salt
Bratwurst	Allspice, caraway seeds, marjoram, white pepper, salt
Breakfast sausage	Sage, ginger, nutmeg, thyme, cayenne pepper, white pepper
Calabrese	Cayenne pepper, fennel seeds, garlic, black pepper, paprika, salt
Chorizo	Onion, garlic, vinegar, red chile, cinnamon, cumin, oregano, salt
Hot dogs	Mustard seed, paprika, mace, celery seed, garlic, black pepper, salt
Italian sausage	Fennel, black pepper, coriander, red pepper flakes, oregano, garlic, sugar, caraway seed, salt
Polish sausage	Sugar, thyme, basil, garlic, mustard seeds, marjoram, salt, pepper

In addition to flavors, there are many textures and binds (course, fine, emulsified, and so on). Some examples of emulsified sausages are hot dogs or mortadella. Coarse-ground sausages are similar to Italian sausage or chorizo, and some examples of finely ground sausages are breakfast sausage or merguez.

A more intimate look at sausage types

"Salumi and charcuterie, Italian and French terms referring to the old practice of full carcass utilization in which curing and preserving are the means to sustain life and extend shelf life, was once a necessity but is now a craft. In modern times, however, *salumi* and *charcuterie* both commonly refer to any meat further processed, both ground and seasoned for consumption. From fresh sausages to more advanced techniques of cooking/smoking, curing, forcemeats, pâtés and terrines, both terms are all encompassing. To gain a basic understanding of the principles of further processing whole muscle and trim into palatable and often unique products, we focus on basic and fresh sausage preparations."

—Mark M. DeNittis of the Rocky Mountain Institute of Meat: Foundations of Meat Fabrication Professional Program Workbook

Types of sausages

Sausage can be made from red meat (beef, pork, lamb, or veal), poultry (turkey or chicken), or a combination. Sausages fall into two general categories: uncooked and ready to eat (RTE).

- ✔ **Uncooked sausage:** Uncooked sausages include fresh (bulk, patties, or links) and smoked sausages. These sausages need to be cooked before eating and require refrigeration because the raw meat in them will spoil.

  To prevent foodborne illness, uncooked sausages that contain ground beef, lamb, or veal should be cooked to 160 degrees. Uncooked sausages that contain ground turkey and chicken should be cooked to 165 degrees. Uncooked pork sausages must be cooked to 160 degrees to prevent illness associated with Trichinella spiralis, a parasite found primarily in pork products.

- ✔ **RTE (ready-to-eat) sausage:** RTE sausage includes sausages that have been cured or processed in a way that eliminates the need for cooking them prior to eating. They also tend to have a longer shelf-life than uncooked sausage. RTE sausage includes meat specialties, as well as smoked/cooked, semi-dry, and dry products. I explain these categories in the next sections.

For detailed information on sausage and food safety, check out the United States Department of Agriculture (USDA) Food Safety and Inspection Services website at `http://www.fsis.usda.gov/factsheets/Sausage_and_Food_Safety/index.asp`.

Meat specialties

A meat specialty is seasoned meat that is chopped, ground, blended, or emulsified (blended at a high enough speed to force the products to bind together) and then hot smoked, cooked, or baked. Meat specialties include products like olive loaf, hot dogs and frankfurters, bologna, and liverwurst. When they're vacuum packed, meat specialties can have a shelf-life of up to 30 days. They also commonly use Cure #1 (salt/nitrite).

Semi-dry sausages

Lots of lunch meats and/or pizza toppings fall into this category. Semi-dry products are usually dried before cooking. They can dry for days or weeks; they can also be poached, baked, and/or smoked. Other examples of semi-dry products include sandwich or pizza pepperoni, Lebanon bologna, summer sausage, and Genoa salumi. These products may also include Cure #1 or an all-natural nitrite alternative.

Cure #1 is a salt and nitrite product that helps the sausage maintain its bright reddish or pinkish color. Without it, the product would be gray or brownish. Today, all-natural nitrites are commonly used, such as celery juice powder and naturally occurring nitrites, in place of sodium nitrite.

Dry sausages

Dry sausages are RTE products that are not cooked but that have been fermented and dried from 21–180 days. They're tested for pH (the pH scale measures how acidic or alkaline a solution is) on a scale after fermentation and water activity before sale. Producers must document and follow a series of rules established by the USDA and the FDA to ensure the safety of these products. A few examples of dry sausages include hard salamis, Portuguese dry linguicia, and Spanish chorico/chorizo.

Dry sausages often use Cure #2 (salt, nitrite, and nitrate) and usually have a moisture content of less than 35 percent of total product. The shelf life of these products is easily several months.

Chapter 15

Sausage-Making Techniques

- -

In This Chapter

▶ Understanding the science behind sausage-making

▶ Familiarizing yourself with the sausage-making process

▶ Tips and suggestions to maximize your sausage-making success

- -

Sausage-making is one of the oldest and most widespread food preservation techniques. It's also an integral part of the butchery process because it enables you to use a great deal of the trim, bits of meat, and fat that are left over from your butchery endeavors.

Understanding basic sausage-making techniques and principles gives you the core knowledge you can later use to replicate any particular quality, texture, or flavor profile that speaks to you. (You can find a variety of sausage recipes in the next chapter that'll get you started.)

In this chapter, you gain an understanding of the key elements to successful sausage-making: I explain the sausage-making process, share some sausage science that can help you create the best-tasting sausage possible, and more.

Getting in Touch with Your Inner Nerd: Sausage Science

To fully understand what makes sausage so wonderful and delicious, you have to get in touch with your inner nerd. You know sausage tastes great, but what really makes it taste so good? Knowing the answer to that question is what separates the novice from the pro.

Sausage is made up of carbohydrates, proteins, fats, and water. Understanding the role those elements play in the end result and their relationship to one another from production to plate is what sausage science is all about.

Producing great-tasting sausage relies on several factors, which I explain in the following sections.

From past to present: Traditional sausage-making

Traditionally — that is, before the days of refrigeration — butchers had two primary objectives: Use as much of the carcass of a slaughtered animal as possible, and find a way to preserve any excess meat so that it didn't spoil. A lot of creative problem-solving and delicious dishes resulted from their efforts, and the history of sausage and preserving is rich in heritage and tradition.

Here's just a glimpse: Whenever an animal was butchered, the blood, offal (internal organs and entrails of an animal that can be used as food), fats, and even the hide and whatever other meat could not be eaten before it spoiled was either salt-cured (in which salt is used to preserve the meat) or made into sausages and dried or smoked. In some cases, the dried and smoked sausages were also cooked and preserved in lard to be stored over the winter; this technique was called *lard packing*. When warm weather came again and the lard softened, it was time to eat the sausages.

In the chilly winter months, low temperatures and salt-curing or lard-packing kept hams, shoulders, and larger portions of the carcass preserved for later use. When the weather warmed up, the meat would be smoked to halt spoilage and provide nourishment in the months to come. Of course, today we have refrigeration, an unlimited a access to ingredients, and modern meat science — all of which have transformed meat curing and sausage-making from a survival necessity to a stage for creativity and technique.

Using quality ingredients

Sausage is no decoy for bad ingredients, so forget *The Jungle,* Upton Sinclair's 1906 novel that was intended to convey the exploitation of immigrant workers in America but ended up horrifying the nation because of its portrayal of sausage-making in a turn-of-the-century Chicago meat-packing plant. Also set aside the notion that "you don't want to know" what goes into the sausage you eat. The fact of the matter is you can't make good sausage without using good ingredients.

"The classic hotdog has such a negative reputation. The reality is it's a profit maker in a shop. Yes, it uses lesser cuts like shank meat, trimmings, and excess fat. Done right, its has stunning and flavorful results. With the exception of salamis, it might be the most difficult 'fresh' sausage there is to make. How's that for irony?" — Bryan Butler, Salt & Time, TX.

Getting the right amount of moisture

Moisture is added to sausage via water or sometimes stocks or vinegars, and should make up at least 10 percent of the recipe. Not having enough liquid in the sausage results in a grainy texture.

Dried spices soak up a lot of moisture; if you're making a chorizo or a sausage with a good amount of paprika in it, for example, make sure you have enough moisture.

Achieving the right texture

The texture of the sausage is determined by the coarseness of the grind, the binding (or emulsification), and the fermentation process, as well as the drying conditions, if applicable. You can experiment with textures by mixing up the ratio of course grind with fine grind (this also helps with binding).

Caramelization and the Maillard reaction are the processes that get the credit for making meat taste so great. In a nutshell, the more surfaces there are to brown, the deeper and richer the flavors become. In 1910, French chemist Lois Camille Maillard observed that there are two unique ways that foods achieve non-enzymatic browning, through caramelization and through the Maillard reaction. The Maillard reaction is a chemical reaction between sugar and amino acid during cooking that generates an elaborate range of smells and flavors within the food. It's the process responsible for complex, palate-pleasing deliciousness. When exposed to heat, foods develop their own unique range of tastes, aromas, and layers of flavor. Flavorists (flavor chemists) use this reaction to engineer natural and artificial flavors for consumption.

Ensuring a good bind

As I note elsewhere, the three main ingredients in sausage are meat, fat, and water. To produce a delicious, juicy sausage, you must properly *bind* the meat by emulsifying it before stuffing.

Emulsification is the glue that holds the sausage together during cooking. During the mixing of meat, seasonings, and fats, salt-soluble proteins are extracted from the meat and coat the fat, creating the bind. When the sausage mixture is cooked, the meat firms up (and the fat and water within your emulsion is suspended in your bind). Without a good bind, the fat and water will drain out of your sausage during cooking, and you'll end up with dry, crumbly sausage. Good emulsification also increases your yield, reduces shrinkage, and gives you the desired texture you're looking for.

Successfully trapping the moisture and rich fats inside the seasoned meat mixture can make or break your sausage. Here are the things you need to know to ensure a good bind:

✔ **Ensuring the right temperatures to make sure the emulsion sets:** Keep the meat and fat as cold as possible when you are mixing. If the mixture is too warm, not only is it not safe, but it can also cause the fat to smear.

✔ **Avoiding the smearing effect:** Smearing happens when a sausage mixture has been over-mixed. You can tell if smearing is happening because the fat will no longer be visible within the mixture as its own entity.

The finer the meat is ground, the less it can be worked. The meat and fat break down too fast and smear. Therefore, try to use finer grinds with simple recipes with few seasonings. When making a courser-ground sausage mixture, you can add lots of seasonings and other ingredients, like whole pepper or dried fruit. Courser grinds can be mixed more before the meat breaks down too much.

✔ **Getting the right lean/fat/water ratio:** The ratio of lean to fat can depend on your personal taste preferences, but a good place to start is 75 percent lean, 25 percent fat. Water should make up around 10 percent of your sausage recipe based on weight. So if you're making 10 pounds of sausage, 1 pound of the total weight should be water.

From Bryan Butler, Salt & Time: "I've found that the finer you grind the meat, the less you mix it; otherwise, you'll develop smearing while stuffing. This is not only visually unappealing, but it also effects the cooking process. Conversely, the courser the meat is ground, the more it can be mixed — the perfect approach for semi-emulsified sausages, like brats."

Using the proper technique

Techniques used during the preservation process, like smoking, curing, or cooking, also affect the consistency, flavor, and balance of the sausage ingredients.

✔ **Smoking:** Smoking is done at or below 200 degrees Fahrenheit. Anything higher cooks the meat with smoke. Ideally you want to take the product to 150 degrees (or the recommended temperature for the animal you're smoking). You can find smoking instructions in Chapter 17.

✔ **Curing:** Curing refers to using salt to preserve meat, either by packing it in a salt mixture or injecting it with a salt solution. For detailed instructions on salt curing meat, head to Chapter 17.

✔ **Cooking:** Some sausages include a pre-cooking step before they are chilled and later grilled. For example, the Beef Bratwurst recipe in Chapter 16 includes such a step. Sausage recipes generally specify what cooking method to use. If they don't, you can cook the sausages in a smoker, with or without smoke. You can also poach or steam them to a safe temperature.

Fermented sausages and guarding against botulism

Fermented sausages are cured sausages; they're classified in two groups: sliceable raw sausages, like salami and pepperoni, and spreadable raw sausages, like Nudja. (Most Americans have never seen spreadable salami, but it's popular in the European Union, and most regions have their own versions.) Fermented meat products are normally consumed without cooking, making them essentially a raw-meat product.

Meat fermentation, much like cheese-making, improves the meat's shelf-life. The combination of fermentation with salting and drying can, in many cases, produce a product that will keep safely at room temperatures. The key danger with fermented sausages, however, is unwittingly creating an environment in which botulism can thrive.

In this book, I do not include recipes for fermented sausages. However, as your sausage-making skills and desires advance, there are some wonderful resources you can use: *Charcuterie: The Craft of Salting, Smoking, and Curing,* by Michael Ruhlman, Brian Polcyn, and Thomas Keller; *Cooking by Hand,* by Paul Bertolli; or *The Art of Making Fermented Sausages,* by Stanley and Adam Marianski are a few of the best references. Just be sure you know what needs to be done to ensure the safety of your final product and that you follow those rules to the letter.

Lowering the pH to inhibit the growth of harmful bacteria

The sausage-maker adds bacteria culture to the meat mixture. This culture, when exposed to the right temperature and humidity, will thrive and consume the carbohydrates in the salami meat mix. The by-product of this activity is lactic acid, which creates a drop in pH, which is a good thing because it keeps other, less desirable or potentially hazardous bacteria at bay, a key first step in making the salami safe for consumption without having to cook it. The fermentation cycle can be 48–72 hours, depended on the specific recommendations of the culture manufacturer.

Guarding against botulism

Salami has yet one more hurdle to cross in order to make it safe. Salami should be air tight, with no oxygen exchange happening. The lack of oxygen is called *anaerobic,* meaning "without oxygen." The anaerobic environment is an integral part of how salami is made, but it also creates the biggest concern when making salami: botulism.

Botulism toxin can occur in any anaerobic environment, like canned foods *and* salami. This demands attention of the sausage-maker (or canner) and must be accounted for. In the olden days, saltpeter was used (which basically in sodium nitrite in very high concentration), but it doesn't disperse well in small amounts. Over time, scientists discovered that it was, in fact, the nitrite that was helping ensure the safety of the meat, and they created the nitrate that is found in Cure #2. Sodium nitrite and nitrates are the most effective way to prevent the botulism toxin and thus ensure the safety of your salami.

You can also find natural celery- or beet-based nitrite cures on the market, but these aren't regulated by USDA or FDA because they can have inconsistent concentrations. "Natural" cures are also harder to find.

Making Sausage: The Basic Steps

The word *sausage* originates from the Latin word *salsicia,* meaning something salted. Technically, sausage is meat that is chopped or minced, seasoned, and then formed into links or tubes. Most people understand sausage to be small (like a breakfast sausage) or medium in size (like a bratwurst), but the term sausage also includes salami and a variety of larger stuffed salumi. Bulk sausage mixtures — even though that aren't formed into any particular shape — are still called "sausage" and are typically sold as-is and formed into patties or sauteed "loose" in a pan and added to pasta sauces, pizzas, and anything else your imagination can possibly pair with spiced and salted meat.

However you define it, sausages are one of the perfect foods. They come in many shapes and sizes and can be made from any protein. Sausages can be sold fresh, cooked, dried, or fermented. They are good hot. They are good cold. They are a great snack or the meal itself. They can be served alone or on/in other preparations. In the following sections, I outline the steps and techniques you need to know before beginning your sausage-making adventures.

Gathering your ingredients

As I note earlier, you want to use the best-quality, freshest ingredients you can find. Remember, the ingredients you start with determine the final outcome. The following sections tell you what you need.

Stephen Pocock, from Boccalone and Damn Fine Bacon, CA, says, "Fat makes the sausage juicy and delectable. If you don't use enough, you end up with dry and crumbly sausage — a disappointment! Don't be afraid of the fat. Fat is good for you, as long as you don't overdo it."

Choosing the best cuts for sausage
Many butchers contend that the best meats for sausage-making tend to be those that come from the forequarter muscles. For pork, that means the

Boston butt; for lamb, the shoulder; and for beef, the chuck. Other butchers prefer different cuts.

The forequarter muscles are often nicely marbled and have great flavor. In addition, the shoulder muscles also tend to have cleaner and richer flavors. (Some of the value cuts from the hindquarter/leg tend to be leaner as well a little iron-y in flavor, due in part to the blood flow during processing that happens "south of the heart.") Pork shoulder is great for sausage because it has almost the perfect ratio of fat to lean, and you can add as much fat as you want or spend some time trimming it out.

Says Bryan Butler from Salt & Time, "On pork butts, the lean-fat ratio fluctuates between pig to pig. I've seen them so fatty, it's 50/50 lean to fat. So never blindly grab shoulders thinking you can always be "right on" for the lean to fat ratio. Always inspect the meat for the qualities you want. If I had to say it, really good pork belly is actually a much more preferable choice, but only if it's cheap and good lean/fat."

Whichever you prefer, always use quality meat from a butcher you trust. Get humanely raised, vegetarian fed (preferably pastured), antibiotic-free meat. Not only is it the right thing to do; it tastes better!

Adding fat

For sausage-making, use fat that is referred to "back-fat" or solid fat. This is the type of fat that comes from along the back of the animal. Belly/bacon fat works well, too.

Don't use subcutaneous fat (soft fats), which tends to be very stringy and is therefore not well suited to sausage-making. Nor do you want to use suet, organ, or caul fat. These fats are too crumbly and won't perform as well when incorporated into a fresh sausage mixture.

Quality spices and the right amount of salt

Make sure you use high-quality spices. Pick those that are fresh and bright. Even if you have to pay more from them, they're well worth the investment because they're more flavorful.

In addition to your spices, you also need salt. Salt is the single most important ingredient to a good sausage. Not only does it enhance the flavor, but it also triggers a reaction (myosin) in protein, making it stickier and easier to bind, a key factor in how good your sausage is. Either too much or too little salt is bad. It's widely accepted that salt should be 2 percent of the total weight. I like to use kosher or California sea salt; black or colored salts are wasted in sausage.

If you use any salty ingredients, like soy sauce, cheese, and so on, adjust the amount of salt in your recipe. In some cases, it could drop it to 1½ percent or less. Conversely, if you use sweeteners, you may need more salt. Always do a taste test to confirm the flavor is balanced.

Going for the cure

Not all sausages require that you use cure. Cure #1 is for smoked or cooked products, like sausages, bolognas, and pâtés. It extends the shelf life of the product and protects its color and flavor.

Cure #2 is used *only* for long-term preservation, like dry-curing and salami-making. Cure #2 contains both nitrite and nitrate. The important thing to know about Cure #2 is that nitrate occurs as the nitrite breaks down during the curing process. it's this that allows the long term drying.

Cure #1 is available at some hunting supply stores. Cure #2 can be bought at specialty meat processing suppliers. You can also find both at butcher supply stores or online retailers.

Stephen Pocock, from Boccalone and Damn Fine Bacon, CA, says, "A note about cure/sodium nitrite: There's been a lot of debate about including sodium nitrite in prepared meat (sausages, bacon, ham, and so on). The cure we use is called a number of things. Two common brands are DQ Cure #1 or Prague Powder #1, but 'pink salt' is the generic term. (It's dyed pink to prevent confusion with regular table salt.) By law, cure is 6.25 percent sodium nitrite and 93.75 percent salt. Used in the proportions suggested here, which are fully compliant with FDA and USDA standards, the risk of getting cancer or any other disease is negligible. So says the American Medical Association. Spinach and celery salts or juice are loaded with nitrate, yet no one is frightened of eating these. Additionally, nitrites are generated naturally. Nitrite is also an antimicrobial. Adding sodium nitrites to cured meats not only gives them a nice pink color but also to makes them safer to eat."

Preparing the meat for grinding

The most important thing to do when you prepare your meat is to get out the parts you don't want to eat and then do as much as you can to separate the lean from the fat in order to control the lean-to-fat ratio. Follow these steps:

1. **Carefully find, remove, and discard any bone or cartilage fragments and glands. Also look for sinews, silver skin, and any other chewy hard bits you don't want in your sausage. Eliminate any loose fat or connective tissue and the outer layer of hard fat.**

2. **Separate the soft fat from the hard fat.**

   The soft fat is great for rendering (head to Chapter 17 for instructions). Save the hard fat for your sausage.

3. **Cut your meat and hard fat into golf-ball–sized pieces or smaller, depending on your grinder.**

   It is easier to attain consistency in the pre-grind preparation in some cuts (hams) than others (shoulder, also called butt).

4. **Assemble your meat and fat according to the ratio set by your recipe.**

   For example, your recipe may want 10 percent of your target weight in hard fat.

   If you're using the same size plate to grind both the lean and the fat, mix the meat and fat together first so that, when you grind it later, the two mix together more evenly.

Chilling the meat before grinding and mixing

One key step in sausage-making is to make sure that your meat and fat mixture is properly chilled. Place it in your freezer until it is nearly frozen (35 degrees). Chilling it to this temperature could take overnight, depending on your batch size. At the least, several hours in the freezer is essential.

Properly chilling the meat-fat mixture is important for a couple of reasons:

✔ **Food safety:** The temperature danger zone, in which nasty things can grow pretty quickly, is from 40–135 degrees. The simple act of grinding and mixing the meat causes it to warm up. Starting with cold meat helps keep it out of the danger zone for as long as possible.

✔ **The bind:** When you grind and mix the sausage, the sausage will warm up. If the meat and fat become too warm, you won't be able to get the bind you want, and you run the risk of the fat breaking down and smearing.

You also want to keep the meat cold during production. To do that, consider freezing the head, auger, blade, and plate grinder components before making sausage.

Grinding and mixing your sausage

Before you begin grinding, ask yourself, "Is the meat and fat mixture cold?" Check it! If it's over 35 degrees Fahrenheit, it's too warm. Place it back in the freezer until it reaches the right chilled temperature. Then you're ready to grind and mix, explained in the following sections.

Grinding your sausage

To grind your sausage, follow these steps:

1. **Assemble your grinder, using the plate specified in the recipe.**

   If you're winging it, choose the plate size that will give you the grind size you like.

2. **Grind away!**

Here are some pointers for getting a good grind:

- ✔ **Let gravity work for you.** Drop the meat into the grinder slowly and try not to force it. Take it slow with smaller table top grinders, because their motors may not be strong enough to push the meat through at an accelerated pace.

- ✔ **Check and make sure all the parts are still tight after you start, especially if you're using a smaller grinder.** Sometimes during grinding, the head may loosen enough for the chopper plate and blades to lose contact with one another.

- ✔ **Look and listen.** If tough sinew or bone passes thru, you'll hear it. Disassemble the grinder and remove the blades, along with whatever bone or sinew got caught up in it. Also watch the meat as it comes out. It should look the way you intended. Lean to fat should be what you expect. If not, stop, evaluate your grind, and adjust as necessary. This is your one chance to get it right.

Mixing your sausage

To mix your sausage, follow these steps:

1. **Set up your mixer if you have one; if not, just use a plastic tub.**

   Refer to Chapter 14 for information on sausage-making equipment, including mixers.

2. **Place your ground meat and fat in the tub and season with your prepared blend of spices, liquids, and cure (if using).**

3. **Use a grasping motion to squeeze the meat and spices together. Get a really good mix.**

   Use latex/rubber gloves! You want to avoid a lot of bare-skin contact with your food, and wearing gloves helps with the really cold meat!

   You want to get a good bind. The bind is what holds the sausage together, whether it's in a casing or just a patty. As you squish everything together (notice how the hard fat keeps its shape? That's because it's so cold!), the lean meat should start to break down and stick together; that's the action of the protein strands meshing. (The same thing happens when you knead dough to make bread.) When you can pick up some of the mixture and pull it apart a wee bit without actually breaking it apart, you've got the bind you're looking for, and you're done.

4. **Make a little patty, sizzle it up, and eat it.**

By doing this little test, you can evaluate the texture, the bind, the effect of the seasonings, and so on and make any necessary adjustments. All these are fixable at this point.

Pondering the big question: Grind first or mix first?

There are two schools of thought regarding which should come first: mixing or grinding. That is, should you mix the seasoning in with the meat first and then grind it, or should you grind it and then mix in the seasoning? Each has its pros and cons, which I outline in the following sections. I prefer to take a pragmatic approach and use whichever method is best for the kind of sausage I'm making.

Method 1: Grinding first

With the "grind first" method, you cut the meat to adequate size (2 inches for average home grinder), grind it through a medium-sized plate, add your seasoning, mix it by hand (like kneading dough) or with a paddle mixer, and then load up the sausage mixture into the stuffer and stuff into casings (explained in the next section).

Here are some things to think about with this method:

- It can lead to a longer processing time.
- The advantage is that you can mix the non-meat ingredients thoroughly and uniformly into the meat-fat mixture.
- The disadvantages include longer handling time, having to use (and clean) multiple pieces of equipment, the stringy long strands of subcutaneous fat from paddle mixing, the friction, and the smearing that can result from over-mixing.

Method 2: Mixing first

In the "mix first" method, you cube the meat depending on the size of the hopper/feed tray (for at-home cooks, that probably means cutting the mean into 2-inch cubes or strips) and dice the fat into slightly smaller pieces for uniform distribution. Then you mix all the spices and liquids with the meat. At that point, you want to let the meat and spices refrigerate for 2–4 hours to allow the flavors to bloom and permeate the meat. But if absolutely necessary, you can grind the sausage and stuff it into casings within 15 minutes of mixing.

Stuffing the sausage into the casing

If you're making link sausage, you'll need casings and a sausage stuffer (refer to Chapter 14 for details on this piece of equipment). You can find casings from different animals (cow, pig, and sheep) and in different sizes: lamb casing, 18-22 mm; hog casing, 28–32 mm and 34–38 mm, for example. Beef middles are a the gold standard for large salami production, and they run 40–50 mm, 2½–3 inches. There is a similar array of synthetic and collagen casings. What you use depends on the recipe but also on what your butcher can get for you. You can also order casings online.

To use a sausage stuffer, follow these steps:

1. **Prep your casings.**

    You need to soak the casings and flush them by running water through them from one end to the other (this removes the salt) before they are ready to slide onto the horn.

2. **Pack your stuffer as tightly as you can and crank the stuffer with a slow, even motion until the meat starts to come out; then stop.**

    When the meat starts to come out, you can slide the casing on the horn. If you put the casing on before the meat has come through, the casing fills with air and blows up like a balloon, and your sausage will have more bubbles in it than Don Ho's champagne.

    As you stuff, try holding the receiving casing slightly higher than the stuffing horn to allow some air to escape.

3. **Load the horn with the casing.**

    Slide as much casing onto the horn as you can. Dribble water over it all, and off you go.

    The enemy in stuffing is air. If a trapped air pocket comes though the horn after you tie off a sausage casing, the sausage will blow apart — literally. Air pockets can also affect how the sausage cooks and how the fats render and pool inside the sausage, which doesn't look too good.

4. **Take your time passing the meat through the horn and into the casing; as you do, be sure to moisten, moisten, moisten.**

    Dry casings tear. By keeping the horn and the casing wet, the casing won't stick to the horn and tear. Don't tie any knots yet; give yourself a few inches of slack.

    Crank evenly and slowly. Don't worry about linking yet. Don't over pack the casing. If you think the casing is sticking, dribble more water on it. As you crank the sausage off the stuffer, you can spiral it into a big coil.

5. **Inspect your work to look for air bubbles; if you find any (and you probably will), use your sausage pricker or a pin to remove them.**

   Air bubbles are almost unavoidable. When you find one, just prick the bubble and gently press out the air.

   At this point, you're ready to link and tie off. Some pros prefer to prick after linking. Doing so puts the sausage under pressure and enabels air to escape more easily. Doing a bit of both is okay. Pressurizing it first really helps; try it sometime.

Tying the knot: Linking and drying sausages

To link sausage, follow these steps:

1. **Start with a good tight square knot or bubble knot.**

   To tie a bubble knot, you tie a piece of twine around the end of the casing, bend the loose edge of the casing so that it doubles back over twine, and then tie it again (see Figure 15-1).

   You can sterilize lengths of twine by microwaving for 30 seconds to 1 minute. You want it above 150 degrees, but you don't want to burn it.

2. **Select the length you'd like and gently pinch the separation at both ends (this is why you don't want to overstuff!).**

3. **Lift this sausage up off the table and spin it away from you about seven times.**

   Be careful not to trap meat inside the twisted end of the casing. Doing so can affect drying.

4. **Measure off the next sausage, pinch off and hold, but *don't spin it!***

5. **Repeat Steps 2 and 3, moving from one sausage to the next and continuing to spin every other sausage link.**

   Alternating spun with unspun links ensures that every other link is counter-spun. You can spin links in the same directions, but if they hang more than three or four down, they typically spin out or uncoil. To stop spin out, try alternating each spun link (this helps, but it can still happen).

6. **Tie another bubble knot at the end.**

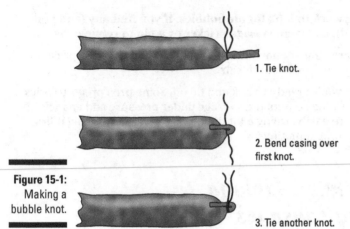

1. Tie knot.

2. Bend casing over first knot.

Figure 15-1: Making a bubble knot.

3. Tie another knot.

Illustration by Wiley, Composition Services Graphics

Hanging your links to dry

One goal of drying is to create a *pellicle,* a tough protective layer that forms when the casing tightens around the sausage. It gives the meat and seasoning a chance to get to know each other. Drying also dries the ends to seal the sausage. If you plan on smoking the sausage, drying is a must. The pellicle allows the smoke to adhere to the sausages during the smoking process, giving you a better, more uniform color.

Drying sausage can be difficult for at-home cooks because hanging the sausage isn't always possible. Dry them inside a refrigerator or sanitary, food-safe environment that is under 50 degrees Fahrenheit.

If you can manage the space to hang your handiwork, I recommend using stainless steel S hooks. Simply hang the hook and drape the sausage from that. If you don't have space to hang your sausage for drying, lay them out flat. Let one side dry and then the flip the links over to let the other side dry. Although this technique won't produce the same uniform color that hanging does, it still does the job.

How long the sausage needs to dry depends on the kind of sausage you're making. In the case of fresh sausage, for example, a couple days of drying usually gets the job done. Leave the links connected during drying to seal the end so that it stays closed during cooking — another sign of a well-made sausage link. After drying, you can separate the links and store them.

To help your sausage links dry faster and more uniformly, follow these pointers:

✔ Clean the outside of the casing of seasonings and any meat that may have gotten stuck to it while you were stuffing.

✔ Give your sausages a quick rinse in water or, better yet, a water-white vinegar mixture (90 percent water to 10 percent white vinegar). Rinsing is a "secret" of the pros.

Storing Your Sausage

To safely and deliciously preserve your creations, you need to store them following a few simple guidelines. See Table 15-1 for a quick review of where and how long you can safely store different kinds of sausage.

Table 15-1	Sausage Storage Chart	
Type of Sausage	*Stored In*	*Storage Time*
Fresh, uncooked sausage	Refrigerator	1–2 days
	Freezer	1–2 months
Fresh, cooked sausage	Refrigerator	3–4 days
	Freezer	2–3 months
Hard/dry sausage	Refrigerator	Indefinitely (whole, unopened); 3 weeks (opened)
	Pantry	6 weeks (whole, unopened)
	Freezer	1–2 months
Hot dogs and other cooked sausages	Refrigerator	2 weeks (unopened); 1 week (opened)
	Freezer	1–2 months
Lunch meat	Refrigerator	2 weeks (unopened); 3–5 days (opened)
	Refrigerator	1–2 months
Semi-dry sausage	Refrigerator	3 months (unopened); 3 weeks (opened)
	Freezer	1–2 months

You may wonder why you can keep certain sausages (like hard or dry sausages) indefinitely in the refrigerator but only 1-2 months in freezer. Here's the answer: The whole process of dry curing for preservation circumvents the need for refrigeration, and freezers are *terrible* environments for meat (at least until you get below 0 degrees, but to get this benefit, the product needs to drop to 0 degrees very rapidly — not something that's possible in your standard-issue home freezer). Freezers are designed to pull moisture out of the air — and as a result, out of anything within it, including meat — which makes frozen sausages dry and crumbly.

Chapter 16

Scrumptious Sausage Recipes

*W*hen you butcher, you invariably produce trim — bits of meat, fat, and skin that you don't otherwise have a use for. Rather than throw this stuff away, make sausage. Sausage-making is a delicious way to use as much of the animal as possible while producing delicious results. In this chapter, I include a variety of recipes — some traditional, some inspired by tradition, and a few just for fun — that help you preserve as much of the meat and trim as possible.

The best tools you can have for sausage making are a gram scale, a grinder, and a sausage stuffer. Refer to Chapters 14 and 15 for information on general sausage-making techniques, equipment, and supplies.

When you're making sausages, I recommend that you use a digital scale, preferably one that can measure in both grams and ounces, because the most accurate and effective unit of measurement for sausage-making is the gram. If you don't have a gram scale, don't worry. I convert the grams into pounds, cups, tablespoons, teaspoons, and so on. Keep in mind, however, that the conversion from grams to other units isn't precise. Here's another recommendation: Buy whole spices, weigh the amounts you need, and then grind them in a spice mill (a miniature coffee grinder just for spices) for the freshest flavor.

Number 1, Number 1, Number 1...

Here's a quick note about cure/sodium nitrite: Sodium nitrite is added to cured meats to not only give them a nice pink color but also to make them safer to eat. Still a lot of debate has occurred regarding the use of nitrite in prepared meats (sausages, bacon, ham, and so on) due to the fear of health risks.

The cure used in some of the recipes in this chapter is called a number of things: DQ Cure #1, Prague Powder #1 (both are brands), pink salt (the generic term), and so on. By law,

these cures are 6.25 percent sodium nitrite and 93.75 percent salt. Used in the proportions suggested here, they are fully compliant with FDA and USDA standards and, according to the American Medical Association, the risk of getting cancer or any other disease from sodium nitrate in these amounts is negligible.

If it makes you feel better, spinach and celery are loaded with nitrate/nitrite, yet no one is frightened of eating these. Additionally, we generate nitrite naturally; it's an antimicrobial.

Chicken and Rabbit Sausage

Chicken and rabbit sausages are mild, delicate, and delicious. Try out these unique recipes and add a little twist to your sausage-making repertoire:

- ✔ **Chicken Breakfast Sausage:** This bulk sausage recipe doesn't need to be stuffed into casings. You can form it into patties for breakfast or even sauté it in a pan, breaking it into pieces for a pasta dish. Delicious!

- ✔ **Chicken Tagine Sausage:** This recipe takes the traditional Moroccan dish of chicken cooked with preserved lemon and olives and turns it into sausage.

- ✔ **Smoky Chicken Jalapeño Sausage:** Spicy and smoky with a little bit of heat, this sausage is a winner.

- ✔ **Rabbit and Sage Sausage with Maple Syrup:** If you like traditional pork breakfast sausage, try it with rabbit. You won't be disappointed.

- ✔ **Rabbit and Wild Mushroom Crepinettes:** Crepinettes are easy to make and wonderful pan seared and served with greens.

Chicken Breakfast Sausage

Prep time: 1½ hour, plus chilling time • **Yield:** 5 pounds bulk sausage

Ingredients	Instructions
2268 grams (5 pounds) deboned chicken legs and thighs, plus extra skin (about 1 pound)	*1* Debone, skin, and remove all the fat from the chicken legs and thighs, keeping the meat separate from the fat and the skin. Check for and remove any leftover cartilage and sinew. Add the extra skin to the skin and fat. Place the meat, skin, and fat in the freezer to chill it to 35 degrees (almost frozen).
30 grams (2 tablespoons) kosher salt	
10 grams (2 teaspoons) fresh ground black pepper	*2* Affix a medium plate to your grinder and grind the chicken meat. Mix the salt in with the ground chicken and refrigerate.
15 grams (1 tablespoon) fresh sage	
15 grams (1 tablespoon) finely chopped fresh thyme	*3* If you have a fine plate (³⁄₁₆ or ⅛ inch), affix it to your grinder and grind the fat and skin. Otherwise, grind the fat and skin twice through the medium plate. Set aside under refrigeration.
15 grams (1 tablespoon) finely chopped fresh rosemary	
5 grams (1 teaspoon) ground nutmeg	*4* In a separate bowl, mix the pepper, sage, thyme, rosemary, nutmeg, paprika, ginger, and the chicken stock together.
5 grams (1 teaspoon) paprika	
2.5 grams (½ teaspoon) ground ginger	*5* Combine the ground chicken and the fat in a large bowl, pour in the liquids and seasonings, and mix well to evenly distribute the fats and seasonings.
237 grams (1 cup) chicken stock or white wine	
	6 Cook a little bit of the breakfast sausage to check the seasoning. If you're happy with the result, place the sausage, covered, in the refrigerate until you are ready to use it.

Tip: Your butcher should be able to set you up with extra chicken skin.

Chicken Tagine Sausage

Prep time: 2 hours, plus chilling time • **Yield:** about 7½ pounds sausage links

Ingredients

2268 grams (5 pounds) chicken thighs

817 grams (1 pound 13 ounces) chicken skin

40 grams (8 teaspoons) salt

91 grams (6 tablespoons) red onion, chopped

36 grams (2 tablespoons plus 1¼ teaspoons) garlic, slivered

3.6 grams (¾ teaspoon) ground ginger

5.4 grams (1 teaspoon) black pepper

5.4 grams (1 teaspoon) chopped parsley

5.4 grams (1 teaspoon) chopped cilantro

2.7 grams (½ teaspoon) white pepper

2.7 grams (½ teaspoon) ground turmeric

1.8 grams (a little less than ½ teaspoon) cinnamon

163 grams (a little less than ¾ cups) preserved lemon rind, chopped

91 grams (6 tablespoons) Kalamata olives, pitted

0.5 grams (a pinch of threads) saffron

10 feet of 22–24mm sheep casings

Instructions

1 Debone and skin the chicken and remove all the fat, keeping the meat separate from the fat and skin. Check for and remove any leftover cartilage and sinew. Add the extra skin to the skin and fat. Place the meat, skin, and fat in the freezer to chill to 35 degrees (almost frozen).

2 Affix a medium plate to your grinder and grind the chicken meat.

3 Add the salt to the chicken and mix thoroughly. Refrigerate this mixture for a few hours, until the meat has broken down and can be squeezed into a sticky paste.

4 If you have a fine plate (³⁄₁₆- or ⅛-inch), affix it to your grinder and grind the fat and skin. Otherwise, grind the fat and skin twice through the medium plate.

5 Place the onion, garlic, and ginger in a food processor and process until you get a fine paste. In a large bowl, mix the onion-garlic paste and the black pepper, parsley, cilantro, white pepper, turmeric, cinnamon, lemon rind, olives, and saffron in with the chicken, skin, and fat until thoroughly combined. Sizzle some up, taste, and adjust the seasonings as needed. Place the sausage mixture in the refrigerator while you prepare the casings and assemble the stuffer.

6 Stuff the casings with the sausage and twist them into links, about 6 inches long.

7 Hang the sausage links overnight in the fridge before cutting them.

Tip: Your butcher should be able to hook you up with the extra skin. Also, to be on the safe side, get a few extra thighs so you can cut in more fat if you want to after the all-important taste test.

Tip: Buy your casings on sleeves, which are handy when you need to slide them on the stuffer horn.

Source: Stephen Pocock, Boccalone and Damn Fine Bacon, CA

Smoky Chicken Jalapeño Sausage

Prep time: 1 hour, plus overnight chilling time • **Yield:** 5½ pounds sausage links

Ingredients	Instructions
1814 grams (4 pounds) chicken, with skin on	**1** Debone the chicken and then cube it. Place the cubed chicken in the freezer until well chilled, almost frozen.
15 grams (1 tablespoon) kosher salt	
454 grams (1 pound) bacon	**2** Affix a medium plate to your grinder and grind the chicken meat. Mix the salt in with the chicken and refrigerate until chilled.
5 grams (1 teaspoon) white pepper	
8 grams (1½ teaspoons) ground caraway seed	**3** Grind the fat, skin, and bacon through the medium plate and refrigerate it until chilled.
8 grams (1½ teaspoons) ground coriander	**4** In a separate bowl, mix together the white pepper, caraway seed, coriander, cilantro, jalapeños, garlic, ice water, and milk powder.
15 grams (1 tablespoon) chopped cilantro	
15 grams (1 tablespoon) minced fresh, seeded jalapeño	**5** Combine the ground chicken and the fat-bacon mixture in a large bowl, pour in the liquid-seasonings mixture, and mix well until the meat, fat, and seasonings are well distributed.
10 grams (2 teaspoons) chopped fresh garlic	
8 grams (1½ cups) ice water	
5 grams (⅓ cup) milk powder	**6** Add the crumbled cheddar cheese and mix to combine. Cook a little bit of the sausage to check the seasoning and adjust as necessary.
227 grams (½ pound) crumbled aged cheddar cheese	
5 to 6 feet of hog casings	**7** Stuff the sausage into the hog casings and then refrigerate overnight before cooking.

Vary It! The bacon in this recipe gives the sausage a little smoky flavor, but for extra smokiness, you can smoke the sausages for about 6 hours at 150 to 180 degrees, or until the internal temperature reaches 145 degrees.

Rabbit and Sage Sausage with Maple Syrup

Prep time: 2 hours, plus chilling time • **Yield:** 5½ to 6 pounds sausage links

Ingredients	Instructions
1814 grams (4 pounds) rabbit meat (2 rabbits, about 2½ pounds each)	**1** Debone the rabbits and cut the meat into ½-inch cubes. Place the meat in the refrigerator. Cube the deboned pork shoulder and place it in the freezer until the meat is chilled to 35 degrees.
1134 grams (2½ pounds) pork shoulder, deboned	
15 grams (1 tablespoon) chopped sage	**2** Mix together the pork, rabbit meat, sage, garlic, white pepper, cumin, and nutmeg.
10 grams (2 teaspoons) chopped garlic	**3** Affix a medium plate to your grinder and grind half of the mixture. Place the other half in a food processor along with ¾ cup of the cream and the maple syrup, and process until the mixture is finely ground. In a large bowl, combine the two batches with the remaining ¾ cup of cream. Mix well, using your hands (wear gloves). Refrigerate until chilled.
5 grams (1 teaspoon) white pepper	
2.5 grams (½ teaspoon) ground cumin	
2.5 grams (½ teaspoon) ground nutmeg	
354.9 grams (1½ cups) heavy cream, divided	**4** Stuff the sausage mixture into the sausage casings and twist to the desired length. If you're making 5- to 6-inch links, you should end up with about 20 to 25 links.
30 grams (2 tablespoons) pure, high-quality maple syrup	
30 grams (2 tablespoons) kosher salt	**5** Refrigerate the sausages overnight.
5 to 6 feet of hog casings	

Tip: Before cooking, make sure to prick the sausages to form pin-sized holes. You can bake, grill, or sear these sausages. Cooking time should be about 15 minutes.

Rabbit and Wild Mushroom Crepinettes

Prep time: 2 hours, plus chilling time • **Yield:** About twenty 4-ounce crepinettes

Ingredients	*Instructions*
1814 grams (4 pounds) rabbit meat (2 rabbits, about 2½ pounds each)	*1* Debone the rabbits and cube the meat into ½ inch cubes. Place the meat in the freezer until it is chilled to 35 degrees. Cube the pork fatback and chill it to 35 degrees as well.
456 grams (1 pound) pork fatback	
15 grams (1 tablespoon) kosher salt	*2* After the meat reaches 35 degrees, combine the pork and the rabbit meat along with the salt, white pepper, black pepper, fennel pollen, lemon thyme, scallions, and chervil. Grind the ingredients through the fine disk of a meat grinder.
5 grams (1 teaspoon) white pepper	
5 grams (1 teaspoon) ground black pepper	
10 grams (2 teaspoons) fennel pollen	*3* In a separate bowl, mix the white wine, chanterelles, and cheese together. Then mix the mushroom-cheese combo into the ground meat until well blended.
15 grams (1 tablespoon) chopped lemon thyme	
15 grams (1 tablespoon) chopped scallions	*4* Form the sausage into patties about 2 inches in diameter and wrap each patty in the caul fat.
15 grams (1 tablespoon) chopped fresh chervil	
59 grams (¼ cup) dry white wine	
118 grams (½ cup) sautéed chanterelle mushrooms, roughly chopped	
45 grams (3 tablespoons) grated Parmesan cheese	
227 grams (½ pound) caul fat	

Tip: This sausage is great served with roasted leeks, watercress, and saba.

Note: Caul fat is the fatty membrane that surrounds the internal organs of some animals like sheep, pigs, and cows. It can be used to wrap roasts or sausage mixtures, holding the stuffing in and adding rich flavor.

Beef Sausage

The all beef sausages in this section are great with mustard and sauerkraut. They're also great on the grill or slowly cooked in a pot of baked beans.

- **Beef Bratwurst:** Traditionally a brat is a classic German pork sausage that is typically grilled or fried. Here I use beef instead of pork.

- **Country Bologna:** This is a great snack on campouts and hiking trips. It also makes the best grilled bologna sandwich when paired with mustard and sharp cheddar.

- **Grandpa Ritz's Sausage:** A classic beef and pork Italian sausage with delicious spice, this version will make you wish you knew Grandpa Ritz personally.

- **Cowboy Braut:** This braut is based on a traditional hot dog formula. Even though it can't be called a dog or frank (because it's not ready-to-eat [RTE] and doesn't contain nitrites), it's a great, all-purpose fresh sausage that's reminiscent of true hot dog flavor.

Beef Bratwurst

Prep time: 4 hours • **Yield:** Fifteen 6-inch sausages

Ingredients	Instructions
2195 grams (4.8 pounds) beef chuck (70 percent meat, 30 percent fat)	*1* Dice the beef the into half-inch cubes. Place the diced beef on a baking tray and freeze for 30 minutes. Place the grinder parts (auger knife and medium plate) in the freezer as well.
48 grams (3 tablespoons) kosher salt	
6 grams (1¼ teaspoon) white pepper	*2* Fit your grinder with the medium grinding plate. Grind the meat one time into a bowl sitting in a bowl full of ice.
4 grams (¾ teaspoon) caraway seed	
2 grams (½ teaspoon) nutmeg	*3* Add the salt, pepper, caraway, nutmeg, ginger, marjoram, eggs, cream, and milk powder to the ground beef and mix well. Grind the mixture one more time through the medium plate into a bowl sitting in a bowl of ice. Mix well, making sure the meat mixture is well incorporated. Place the sausage mixture in the refrigerator until you're ready to stuff.
1 gram (¼ teaspoon) ginger	
2 grams (½ teaspoon) marjoram	
4 eggs	
91 grams (6 tablespoons) heavy cream	
91 grams (6 tablespoons) dry milk powder	*4* Using a medium stuffing tube, stuff the sausage mixture into the sausage casings and twist to form 6-inch links. Place the linked sausages in the refrigerator until ready to cook.
6 to 10 feet of natural pork casings, soaking in water	
	5 Preheat the oven to 200 degrees. Cut the links apart and place them on baking trays 1 inch apart. Cook the links for 45 minutes, or until their internal temperature is 150 degrees.
	6 Remove the links from the oven and cool to room temperature. The sausages are now ready for the grill.

Tip: If you don't plan to eat the sausages right away, store them in the refrigerator. They'll be even more delicious!

Source: *Craig Deihl, Cypress, SC*

Country Bologna

Prep time: 2–3 hours, plus overnight refrigeration • **Cook time:** 3 hours • **Yield:** Seven 10-inch bolognas

Ingredients	Instructions
3000 grams (6.5 pounds) beef chuck 163 grams (½ pound) beef fat 15 grams (1 tablespoon) ground black pepper 12 grams (2½ teaspoons) ground coriander 4 grams (1 teaspoon) garlic, crushed 50 grams (3⅓ tablespoons) salt 113 grams (½ cup) ice water 70 grams (4½ tablespoons) light brown sugar 8 grams (1½ teaspoons) Cure #1 15 to 20 feet of beef middles, soaking in water	**1** Dice the beef and beef fat the into ½-inch cubes. Mix in the pepper, coriander, garlic, and salt. Place the mixture on a baking tray in the freezer for 30 minutes. Place grinder parts in the freezer, as well. **2** Using the fine grinding plate, grind the meat one time into a bowl sitting in a bowl of ice. **3** Add the ice water, brown sugar, and Cure #1 to the ground mixture and mix well. **4** Grind the mixture one more time through the medium plate into a bowl sitting in a bowl full of ice. Mix well, making sure the ingredients are well incorporated. Place the sausage mixture in the refrigerator until you're ready to stuff it. **5** Place the beef middle on the stuffing tube and stuff the mixture as tightly as possible into the beef middle, making 12-inch long links, tying twine knots between the links. Refrigerate the sausage while you clean up. **6** Set your smoker to 180 degrees with full smoke, and smoke the sausages for 3 hours, or until they reach 160 degrees. When the sausages are done, place them in the ice water to rapidly chill them. When they're cold, transfer them to a baking tray fitted with a resting rack. Place the sausages in the refrigerator for one day before eating them.

Tip: To secure the twine at the end of each link, tie a knot at the end of the casing, loop the casing over the knot and tie again, loop the string over one more time and tie under the first two knots and finish with another knot.

Source: *Craig Deihl, Cypress, SC*

Grandpa Ritz's Sausage

Prep time: 2 hours, plus overnight refrigeration • **Yield:** 5½ pounds sausage links

Ingredients	Instructions
1814 grams (4 pounds) beef brisket, trimmed of fat and diced	**1** Combine the beef brisket, pork butt, and pork fatback. Refrigerate until you're ready to grind.
454 grams (1 pound) pork butt, diced	**2** Using the fine plate, grind the meat and fat. Put the mixture in the freezer for a half hour.
227 grams (½ pound) pork fatback, diced	
½ head of garlic, puréed	**3** Mix the garlic with the wine and place in the refrigerator for about an hour.
473 grams (2 cups) red or white wine, chilled	**4** Add the salt, sugar, pepper, nutmeg, cloves, allspice, cinnamon, and the garlic-wine concoction to the meat mixture. Blend thoroughly by hand.
Kosher salt to taste	
15 grams (1 tablespoon) sugar	**5** Stuff the sausage into the casing, twisting to form 6-inch links.
30 grams (2 tablespoons) black pepper	
5 grams (1 teaspoon) nutmeg	
5 grams (1 teaspoon) ground cloves	
10 grams (2 teaspoons) ground allspice	
10 grams (2 teaspoons) ground cinnamon	
12 feet of hog casings, 32 mm diameter, stored in water, then soaked in hot water prior to use	

Source: *Marissa Guggiana, author of* Primal Cuts: Cooking with America's Best Butchers, *CA*

Cowboy Braut

Prep time: 2 hours • **Yield:** 5 pounds sausage links

Ingredients

1678 grams (3½ pounds) beef chuck

472 grams (1 pound) pork fat

68 grams (4½ tablespoons) water

68 grams (4½ tablespoons) powdered milk

36 grams (2½ tablespoons) ground English mustard

27 grams (2 tablespoons) salt

14 grams (1 tablespoon) dextrose

14 grams (1 tablespoon) Spanish paprika

4.5 grams (1 teaspoon) ground mace

4.5 grams (1 teaspoon) garlic, granulated

4.5 grams (1 teaspoon) finely ground black pepper

4.5 grams (1 teaspoon) finely ground white pepper

1.8 grams (⅓ teaspoon) celery seed

6 feet of 32 mm collagen casing or 32 mm of natural hog casing

Instructions

1 Dice up the beef chuck and pork fat.

2 Add the water, powdered milk, mustard, salt, dextrose, paprika, mace, garlic, black and white pepper, and celery seed to the meat-fat mixture.

3 Grind the ingredients through a medium plate into stuffing horn and casing, making 4- or 5- inch links.

Vary It! For a slightly different flavor use sweet or hot smoked paprika in place of the Spanish paprika. The subtle differences in the paprika flavor can give the dogs a sweet or spicy twist.

Note: If you want these brauts to mimic that ballpark hot dog snap, go for the collagen casing.

Source: *Mark DeNittis, Il Mondo Vecchio Salumi, CO*

Pork Sausage

Pork. My favorite. Very versatile, and the fat from pork is perfect for making sausage. If you're making lamb, goat, or venison sausage, you can even use pork fat to help juice it up!

Here are the recipes you'll find in this section:

- ✔ **Fresh Fennel Sausage:** This sausage is great on its own or in a bun with peppers and onions. You can also serve it with white beans and kale.

- ✔ **Chaurice:** Chaurice is a spicy pork sausage with Creole roots and Spanish origins that's often served as a side dish with beans. The chaurice recipe included here doesn't contain the traditional celery and green onions but is simple and great.

- ✔ **Ciccioli:** A classic style of preparation from Emilia Romagna, the simple definition is food prepared in pig fat.

- ✔ **Spicy Southwest Sausage:** If you love barbecue and smoky flavors, you'll love Southwest-style sausage, which gets some smoke from chipotle.

- ✔ **Brad's Black Pudding:** This black pudding is a simple sausage with training wheels: It's easy to make yet yields heaps of praise from those who allow themselves to indulge in its rich deliciousness. Simply mix all the ingredients together, stir (preferably with your hands to feel your way through any large chunks of meat or fat), and pour into a terrine mold (or bread tin) to slowly cook in a hot water bath.

- ✔ **Farmstead, Inc.'s Headcheese:** To make a perfect headcheese, you boil down a pork head in herbs and a celery-onion-carrot mixture (called a *miripoix),* pull the meat from the head, and reserve it in the reduced gelatinous stock.

Fresh Fennel Sausage

Prep time: 2 hours, plus overnight refrigeration • **Yield:** Fifteen 6-inch sausages

Ingredients	Instructions
3000 grams (6.6 pounds) pork shoulder (60 percent meat, 40 percent fat)	**1** Dice the pork shoulder into ½-inch cubes. Place it in a large bowl and add the salt, sugar, fresh fennel, ground and whole fennel seeds, garlic, white pepper, red pepper, and cayenne pepper. Mix well. Place the sausage mixture on a baking tray and place in the freezer for 30 minutes. Place the grinder parts (auger knife and medium plate) in the freezer, as well.
48 grams (3¼ tablespoons) kosher salt	
15 grams (1 tablespoon) sugar	
60 grams (4 tablespoons) fresh fennel, chopped fine	**2** Fit your grinder with the medium grinding plate and grind the meat into a bowl sitting in a bowl of ice. Mix well, making sure that the meat mixture is well incorporated; then place the sausage in the refrigerator until you're ready to stuff.
4 grams (¾ teaspoon) ground fennel seed	
2 grams (½ teaspoon) fennel seed, whole	
10 grams (2 teaspoons) garlic, crushed	**3** Stuff the sausage into the casing. Then twist to form the sausages into 6-inch links. Place the linked sausages in the refrigerator for 1 day. When you're ready to cook the sausages, clip them into individual links.
4 grams (¾ teaspoon) ground white pepper	
3 grams (½ teaspoon) crushed red pepper	**4** To cook the sausage, coat the bottom of a large pan with oil and place over medium heat. Place as many sausages as you can in the pan, leaving a half inch between them. Cook on the first side for 4 to 5 minutes, turn them over, and then cook for another 4 to 5 minutes, or until their internal temperature reaches 155 degrees. Remove the links from pan and let rest on a plate for 4 minutes.
1 gram (¼ teaspoon) cayenne pepper	
10 feet of hog casings	

Source: Craig Deihl, Cypress, SC

Chaurice

Prep time: 2 hours 15 minutes • **Yield:** 5 pounds chaurice

Ingredients	*Instructions*
2268 grams (5 pounds) pork butt, untrimmed, boneless	*1* Dice up all the meat ingredients into 1-inch cubes.
45 grams (3 tablespoons) kosher salt	*2* Place the cubed meat into a large bowl and add the salt, onion flakes, parsley, thyme, cayenne pepper, garlic, red pepper flakes, allspice, bay leaves, and water. Mix well.
78 grams (⅓ cup) dried onion flakes	
26 grams (5 ¼ teaspoons) dried parsley	*3* Grind the mixture through a medium plate into the stuffing horn and casing, making 4- to 5-inch links.
10 grams (2 teaspoons) dried thyme leaves	
10 grams (2 teaspoons) cayenne pepper	
5 grams (1 teaspoon) garlic, granulated	
5 grams (1 teaspoon) red pepper flakes	
2.5 grams (½ teaspoon) ground allspice	
2 bay leaves, ground	
113 grams (½ cup) water	
6 to 10 feet of hog casings	

Source: *Mark DeNittis, Il Mondo Vecchio Salumi, CO*

Ciccioli

Prep time: 4 hour, plus overnight refrigeration • **Yield:** 1 terrine (about 20 slices)

Ingredients	Instructions
1135 grams (2½ pounds) pork shoulder, meat and fat separated, cut into 2-inch cubes	**1** Combine the pork, salt, cure (pink salt), and bay leaf in a pot just large enough to hold the meat. Cover with pork stock. Bring to a boil; then reduce the heat and simmer for 3 hours.
570 grams (2½ cups) pork stock, as needed	**2** Strain the meat from the stock and reserve the broth. Chop the meat very coarsely.
17 grams (2½ teaspoons) kosher salt	
2.8 grams (½ teaspoon) pink salt (optional)	**3** Combine the dry spices and grind fine. Add the spices to the chopped meat. Adjust the seasoning with additional kosher salt (you want to overseason slightly because the finished product will be served cold).
1 bay leaf	
2.5 grams (½ teaspoon) mustard seeds, ground	
5 grams (1 teaspoon) coriander seeds	**4** Line a loaf pan or terrine mold with plastic wrap. Gradually fill the mold with the meat mixture, adding the reserved stock to cover it as you go.
5 grams (1 teaspoon) fennel seeds	**5** Cover the terrine with plastic wrap, weight the top, and place it in the refrigerator overnight.
5 grams (1 teaspoon) black peppercorn	
2.5 grams (½ teaspoon) dried thyme	**6** To serve, slice the ciccioli into slices a little thinner than 1 centimeter. Serve with grilled bread.

Source: Matthew Jennings, Farmstead, Inc.

Spicy Southwest Sausage

Prep time: 1½ hours • **Cook time:** 15 minutes • **Yield:** 5 pounds sausage links

Ingredients	*Instructions*
2057 grams (4½ pounds) pork shoulder 229 grams (½ pound) pork fat 32.5 grams (2 tablespoons) salt 10 grams (2 teaspoons) black pepper 10 grams (2 teaspoons) sugar 9 grams (2 teaspoons) ancho chili powder 1.4 grams (⅓ teaspoon) Cure #1 4 grams (½ teaspoon) dried thyme leaves 10 grams (2 teaspoons) garlic, slivered 5 grams (½ teaspoon) chipotle in adobo, chopped 6 feet of 33–35mm hog casings	*1* Cut the shoulder meat and fat into golf ball-sized pieces and mix them together. Then place them in the freezer until they reach 35 degrees. *2* In one small bowl, mix together the salt, pepper, sugar, chili powder, and cure. Set aside. In another small bowl, mix together the thyme, garlic, and chipotle. *3* When your meat has reached 35 degrees, grind it, using the medium plate. *4* Mix the dry and wet ingredients into the meat, using your hands to get a great bind. *5* Sizzle a little of the sausage up, taste it, and adjust as necessary. (*Note:* If the seasoning is too strong, return the sausage to the fridge while you grind up a little more meat; then add the freshly ground meat to the sausage mixture.) Return the mixture to the refrigerator while you prepare your casings and assemble the stuffer. *6* Stuff the sausage and twist them into 5-inch links. Hang the links in the fridge overnight.

Tip: These sausages are great sizzled up with some huevos ranchero and tortillas for breakfast!

Source: Stephen Pocock, Boccalone and Damn Fine Bacon, CA

Farmstead, Inc.'s Headcheese

Prep time: 7–8 hours, plus overnight soaking and ripening time • **Yield:** 1 terrine

Ingredients	Instructions
1 pig head (about 8 to 11 pounds)	**1** Place the pig head in a large container and cover it with cold water. Let the head sit in the refrigerator overnight. Drain.
299 grams (⅔ pound) carrot, chopped	
299 grams (⅔ pound) onion, chopped	**2** Place the pig head into a large pot. Cover it with enough cold water to submerge the head under 6 inches of water. Wrap the carrot, celery, and onion in cheesecloth and tie with twine. Add the vegetables, seasoning sachet, and wine to the pot.
299 grams (⅔ pound) celery, chopped	
1 small seasoning sachet, a combination of black peppercorns, coriander seed, fennel seeds, mustard seeds, garlic cloves, tarragon leaves, thyme sprigs, and juniper berries	**3** Bring the head and other ingredients to a boil, skimming off any scum that rises to the surface. Lower the heat to simmer and cook, uncovered, for 6 to 8 hours. Strain, saving both the liquid and the head, but discarding everything else.
1 bottle (750 ml) dry white wine	**4** Return the cooking liquid to a pot and cook over low-medium heat for a couple of hours, until the liquid has reduced by about half, skimming as needed.
45 grams (3 tablespoons) fresh herbs, such as parsley, chives, sage, rosemary, lemon thyme, and chervil	**5** When the head is cool enough to handle but still warm, pick the meat from the skull. Squeeze the fat into small pieces and pull the flesh into strips. Transfer the meat to a bowl and season with the fresh herbs, salt, black (or Espelette) pepper, and vinegar.
Salt to taste	
Pepper (or Piment d' Espelette) to taste	
Vinegar (such as red wine, champagne, or balsamic) to taste	**6** Add ½ cup of the reduced poaching liquid to the bowl and mix with a spoon until the mixture resembles a soft cheese spread.
	7 Pour the mixture into a terrine mold or loaf pan lined with plastic wrap. Cover with plastic wrap, place a weight on top of the mold, and let the mold set overnight in the refrigerator; then let it ripen for at least 2 days or up to 1 month before serving. Serve with mustard and salt.

Source: *Matthew Jennings, Farmstead, Inc.*

Fried Pork Skins (Chicharrones)

Chicharrones, or *chicharra,* are crispy, dehydrated pork skins that are fried, seasoned, and eaten as tasty snacks. They can be sweet or savory, used for dipping, or eaten alone. They can also be stewed in a salsa-like broth, red or green, and eaten with tortillas. But when fried, the perfect, light, airy crunch of a properly prepared chicharrone is quite a treat. So don't waste those pork skins!

To make chicharrones, follow these steps:

1. **Gather your ingredients: 1 pound of pork skin and your favorite seasoning blend.**

   You can use as much or as little pork skin as you have available, but I recommend starting with at least 1 pound.

2. **Allow your pork skin to dry out by leaving it unwrapped in the refrigerator for a couple days.**

   It should be dry and not easily cut.

3. **Scrape the fat from the skin and cut the skin into to small pieces with scissors.**

   Make sure you clean all the tissue from the inside of the skin. If you don't, you won't be able to produce light, airy, crispy chicharrones.

4. **Boil the skin in a large pot, just until the remaining tissues come off the skin with light scraping, about 45 minutes.**

Boiling helps clean and further remove fat and tissue from inside of skin, but you don't want to boil too long.

5. **Dehydrate the skins at 100 degrees for 12 hours.**

   If you have a dehydrator, great! If not, try laying the skins out on a sheet pan, leaving it in a warm place or in your oven overnight.

6. **When the pork skin is ready to cook, add peanut oil to a frying pan and heat to 400 degrees.**

   The higher the heat, the better because you want the dehydrated skins to cook very fast. If you don't have peanut oil, use another oil that has a high smoke point.

7. **Fry the dehydrated pork skins for 5 to 10 seconds, or until they "pop," and then transfer them to a large bowl and toss with your seasoning mixture.**

8. **Enjoy!**

Tip: Don't have a favorite seasoning blend? Try one of these: Paprika, garlic powder, and salt in 2:1:1 ratio; or go sweet with cinnamon and brown sugar in 1:2 ratio.

Source: Bryan Butler, Salt & Time, TX

Brad's Black Pudding

Prep time: 2½ hours • **Yield:** 2 terrines (about 40 slices)

Ingredients	Instructions
473 grams (2 cups) chopped onions	**1** Combine the onions, garlic, and duck fat in a sauté pan and cook over low heat until very soft. Raise the heat to high and, when pan is hot, add the port. Cook off the alcohol and continue cooking until almost dry. Remove the pan from the heat and cool the onion-garlic mixture completely.
6 cloves garlic, minced	
30 grams (2 tablespoons) duck fat or butter	
50 grams (a little less than ¼ cup) port	**2** While the onion-garlic mixture cools, toast the fennel seeds, coriander seeds, mustard seeds, and cumin seeds. Grind them together with the bay leaf and Aleppo. Set aside.
2.5 grams (½ teaspoon) fennel seeds	
5 grams (1 teaspoon) coriander seeds	**3** In a large bowl, whisk the eggs and add the onion-garlic mixture, the toasted seed mixture, the ground venison or pork, the back fat, the pork blood, and the remaining ingredients.
10 grams (2 teaspoons) white mustard seeds	
5 grams (1 teaspoon) cumin seeds	**4** Break up the ingredients with your hand (use gloves) and stir for 8 to 10 minutes to make sure everything is well combined and thickened.
1 bay leaf, crumbled	
5 grams (1 teaspoon) Aleppo (sun-dried Syrian chili pepper)	**5** Preheat the oven to 325 degrees. Line two terrine molds with plastic wrap, leaving enough to wrap the top after the terrines are filled. Place the pudding mixture in the lined terrine molds and fold the plastic wrap over the top. Cover the top of the terrines with three layers of foil and place them in a hotel pan or casserole dish.
2 eggs	
680 grams (1½ pounds) ground venison or pork	
250 grams (½ pound) back fat, diced	
760 grams (3 cups) pork blood	
177 grams (¾ cup) cooked rice	
177 grams (¾ cup) cooked barley	

118 grams (½ cup) oats

118 grams (½ cup) chopped
fresh parsley

5 grams (1 teaspoon) chopped
fresh sage

5 grams (1 teaspoon) chopped
fresh thyme

5 grams (1 teaspoon) chopped
fresh oregano

2.5 grams (½ teaspoon) ground
allspice

5 grams (1 teaspoon) curry
sausage spice or mild curry
powder

10 grams (2 teaspoons) ground
black pepper

15 grams (1 tablespoon)
smoked paprika

30 grams (2 tablespoons) salt

6 Fill the hotel pan with enough boiling water to reach half way up the side of the terrines. Cover the whole hotel pan with foil and bake until the internal temperature of the pudding reaches 155 degrees, about 1 hour.

7 Refrigerate the pudding overnight.

8 To serve, slice the black pudding into ½-inch thick pieces and crisp them in a pan with hot duck fat.

Note: After you assemble the laundry list of interesting and eclectic ingredients, black pudding is one of the easiest sausages to make. In addition, black pudding freezes well, which is good because this recipe makes more than enough.

Source: Brad Farmerie, PUBLIC, NY City

Lamb and Goat Sausage

Making lamb and goat sausages is a wonderful way to use the trim from the butchering process. They have a rich, distinct flavor and are well-suited for winter months and hearty flavors like rosemary or roasted peppers.

- ✔ **Lamb Kokoretsi:** This fresh sausage recipe uses the same basis as the classic Greek offal-entrails sausage with a few changes. Rather than wrapping the sausage mixture in organ fat and then trussing with intestine, it uses a larger diameter (28-30 mm) sheep casing. Either way this simple, rich, and classic dish can be served with simple tzatziki, hummus, tabouleh, and warm sumac dusted pita chips.

- ✔ **Lamb Merguez:** Merguez is a traditional North African sausage made with lamb or lamb and beef. You can use this tasty sausage for grilling, in meatballs, or crumbled in dishes.

- ✔ **Lamb with Rosemary and Black Olives:** Rosemary and black olives are a classic savory and salty combination, and this sausage highlights the blends of flavors beautifully. This sausage is great grilled or served with watercress and whole grain mustard.

Lamb Kokoretsi

Prep time: 1½ hours • **Cook time:** 15 minutes • **Yield:** 5 pounds sausage links

Ingredients	Instructions
680 grams (1½ pounds) lamb shoulder or 80/20 ground lamb	**1** Dice up the lamb shoulder (if you're using it instead of ground lamb), heart, kidney, liver, and fat. Place in a large bowl.
454 grams (1 pound) lamb heart	
454 grams (1 pound) kidney	**2** Add the salt, onion flakes, mint, oregano, pepper, and garlic to the ingredients in the bowl. Mix well.
454 grams (1 pound) liver	
227 grams (½ pound) lamb or pork fat	**3** Grind the mixture through a medium plate into the stuffing horn and casing. Twist to form 4- to 5-inch links.
45 grams (3 tablespoons) kosher salt	
59 grams (¼ cup) onion flakes, dry	
25 grams (1¾ tablespoons) mint, dry	
5 grams (1 teaspoon) oregano, dry	
5 grams (1 teaspoon) ground black pepper	
5 grams (1 teaspoon) granulated garlic	
6 to 10 feet of 28–30 mm sheep casings	

Vary It! Instead of salt, mint, and oregano, substitute the Za'tar blend (which includes sun dried herb, salt, sesame seed, and sumac) that is sold in most Mediterranean markets.

Source: Mark DeNittis, Il Mondo Vecchio Salumi, CO

Lamb Merguez

Prep time: 1½ hours, plus overnight refrigeration • **Yield:** 5 pounds sausage links

Ingredients	Instructions
118 grams (½ cup) Harissa (see the following recipe)	**1** Prepare the Harissa (see the following recipe). Set aside until ready to use.
1134 grams (2½ pounds) 70/30 ground chuck	**2** Break up the ground chuck and lamb into a large bowl and evenly distribute, but don't mix yet.
1134 grams (2½ pounds) ground lamb	
6 cloves fresh garlic, minced into a paste	**3** Gently massage the garlic paste into the meat mixture.
45 grams (3 tablespoons) dried onion	**4** Mix the onion, pepper, cumin, oregano, and coriander seeds into the sausage mixture. Add the Harissa and mix.
30 grams (2 tablespoons) cracked pepper	
30 grams (2 tablespoons) cumin	**5** Add the cold water to the sausage mixture and blend until the sausage is pasty and mixes easily.
15 grams (1 tablespoon) fresh, roughly chopped oregano	**6** Stuff into a lamb casing and twist to form 5- or 6-inch links. Hang overnight in the refrigerator to dry.
10 grams (2 teaspoons) coriander seeds	
591 grams (2½ cups) ice cold water	
22.5 grams (1½ tablespoons) kosher salt	
6 to 10 feet of lamb casings	

Harissa

Ingredients	Instructions
15 or so medium-sized cayenne chilies	**1** Combine the chilies, salt, cumin, coriander, caraway seeds, and garlic in a food processor and give it a few pulses until it's finely chopped.
10 grams (2 teaspoons) sea salt, or to taste	
10 grams (2 teaspoons) ground cumin	**2** Add the oil and juice, and pulse a few more times until you get a stiff paste that doesn't easily settle. (You want a concentrated paste, not watery or thin.) Set aside until ready to use.
10 grams (2 teaspoons) ground coriander	
2.5 grams (½ teaspoon) caraway seeds	
5 cloves of fresh garlic, minced to a paste	
30 to 40 grams (3 to 4 tablespoons) of quality cooking oil (olive, grape, or vegetable)	
Juice of 1 Meyer lemon to taste	

Tip: While mixing the sausage ingredients together, keep the meat cold by working in a pre-chilled bowl.

Note: The key ingredient in merguez is a hot chili paste known as Harrisa. One thing about Harrisa is that interpretations of it vary from person to person. Feel free to add or take away any particular ingredient and add your own touch.

Tip: For a fresher flavor for your Harissa, try toasting and grinding cumin and coriander seeds rather than using ground from a bottle.

Tip: Mincing garlic to a paste helps to better incorporate the flavor because there's no settling or chunks. An easy way get the paste is to use the side of your chef's knife to smash the garlic.

Source: Bryan Butler, Salt & Time, TX

Lamb with Rosemary and Black Olives

Prep time: 1½ hours • **Yield:** 6 pounds sausage links

Ingredients	Instructions
2268 grams (5 pounds) lean lamb meat or leg of lamb 454 grams (1 pound) lamb fat	**1** Cube the lamb meat and fat into 1-inch pieces. Place them in the refrigerator for a couple of hours or in the freezer about 30 minutes, until chilled.
59 grams (¼ cup) red wine 29.5 grams (⅛ cup) chopped black olives (not canned), kalamata olives, or oil-cured black olives, plus another ⅛ cup if needed	**2** Grind the meat and fat through the fine disk of a meat grinder. Refrigerate for another 30 minutes or so.
10 grams (2 teaspoons) chopped garlic 5 grams (1 teaspoon) whole yellow mustard seeds	**3** In another bowl, combine the wine, half the olives, the garlic, mustard seeds, black and white peppers, salt, and rosemary. (*Note:* Because olives vary in salt content, add half the olives to start. After the meat is mixed with the wine and seasonings, you can always add more.)
5 grams (1 teaspoon) fresh ground black pepper 5 grams (1 teaspoon) white pepper	**4** Using your hands, mix the chilled ground meat and wine-seasoning mixture until well combined. Fry up a little bit and taste, adjusting (adding more olives, for example) as necessary.
15 grams (1 tablespoon) kosher salt 15 grams (1 tablespoon) fresh rosemary leaves, chopped 5 to 6 feet of lamb or hog casings	**5** Stuff the filling into the casing, prick the sausages to release the air pockets, and twist into 4- to 5-inch links.

Chapter 17

Processing Techniques: The Good Kind

. . . .

In This Chapter

▶ Curing and smoking meat

▶ Creating scrumptious dishes with leftovers from the butchery process

▶ Using the bones for stocks and gravies

▶ Rendering fat

. . . .

*T*he term "processing" in artisinal butchery means the butchery of an animal. "Further processing" refers to curing, stuffing, cooking, and smoking all the leftover bits of the animal. This timeless, beautiful extension of the butchery craft produces an array of exciting, delectable products. So forget the less-than-appetizing images of chemicals and preservatives that the phrase "processed food" conjures up. To butchers, the phrase means freshly smoked bacon, a year-old ham salted and aged to perfection, or a perfect duck leg confit, rich with flavor. The information in this chapter gives you the framework you need to process your meat deliciously further.

Whole-Muscle Curing

As a general culinary term, *curing* is a process through which food is dried, smoked, or salted in order to preserve it. In terms of meat, *curing* means to preserve the meat with salt and nitrites/nitrates.

You can achieve successful cured meat via a variety of curing methods:

▶ **Dry curing:** In dry curing, you pack the meat in salt-nitrite mixture. During the aging process, flavors develop as salt is taken into the meat and moisture is expelled, drying and preserving the meat.

▶ **Wet curing:** In wet curing, you soak the meat in brine (a salt and water mixture) or inject/pump a brine solution into the meat. During the wet curing process, salt is taken into the meat while water inside the meat is expelled.

✔ **Combination curing:** Combination curing is a mixture of dry and wet curing.

Several factors determine which of these methods you use: the size of the piece you are curing, the temperature and humidity of your curing environment, how much fat is in the meat, and the meat's moisture content. These factors determine how much curing solution you use.

Following the general process

The process from start to finish involves first curing the meat and then hanging (drying) the meat under controlled temperature and conditions. To begin whole-muscle curing, I recommend using the ham (pork leg). It's a lean, large muscle system made up of many smaller muscle groups that are great for curing. In a nutshell, here's the process you'll follow:

1. **Prepare the meat for curing.**

   Seam out the leg and isolate the smaller muscle groups. Trim any soft fats, visible veins, or blood spots from the meat.

2. **Coat the meat with your curing mixture.**

   • **If you're dry curing,** add your dry cure mixture (salt, nitrites, and seasoning) to the outside of the meat.

   • **If you're wet curing,** immerse the cut in the brine solution.

   • **If you're combination curing,** you use both of these methods (brining and "salting").

3. **Store the meat in a temperature-controlled location until the curing is complete.**

Follow these guidelines when curing meat:

✔ **For dry curing:** How long a whole muscle needs to cure before hanging depends on the size of the cut, either in size of cut or thickness:

   • **By weight:** Use this chart if you're going on size:

Size of Cut	Curing Time
Smaller cuts (bellies, loins)	2 days per pound
Larger cuts (hams, shoulders)	3 days per pound

   Following this method, a 5-pound loin requires 10 days of cure, and a 10-pound shoulder needs 30 days.

- **By thickness:** One formula recommends 7 days of curing per each inch of thickness. Measure the meat at its thickest point. A ham weighing 10 pounds that's 4 inches thick (through the thickest part), for example, would need to be cured for 28 days ($4 \times 7 = 28$).

✔ **For wet curing:** Smaller cuts of meat need anywhere from 3 to 14 days in the wet cure at 40 degrees Fahrenheit. You must turn the meat daily while it is in the brine solution and make sure that the cut is fully submerged in the brine throughout the entire process. Also, scrape away any foam or scum that floats to the top during the curing time.

Although wet curing is a slower process, a small percentage of weight is gained during. After wet curing is completed, you must rinse the meat well and allow the surface of the meat to dry completely before smoking.

✔ **For combination curing:** You use both immersion and dry curing. For example,v a meat cut can be spray pumped or rubbed with a curing solution before being fully immersed in a container of brine. The cuts should be weighed down and turned at least once a day during the entire curing process. Combination curing is a popular processing method because it is quick and cuts down on curing time.

Sounds simple enough, right? Well, for the most part it is, but there's more to curing than just slapping some salt on a pork leg and letting it sit in your refrigerator for a while. Finding the right balance between time (how long to let the meat cure) and temperature is where the mastery comes in to play. Crafting a formula and a product you can be proud of takes practice and patience. In the next sections, I tell you what equipment you need, explain the different types of cure and how to make your own, and share the safety precautions you need to follow.

This explanation is a very quick (and I do mean quick) overview of whole-muscle curing. Meat curing is a science that takes into consideration many direct and environmental variables. Feel free to do your own research and find more comprehensive information on the matter.

Identifying the equipment you need

Some basic equipment is required if you want to begin whole-muscle curing at home. The most important is a place to store and hang your meat at a controlled temperature. Here are some of your options:

✔ Purchase a small refrigerator (make sure you have a thermometer to monitor the temperature) or counter-top humidor.

✔ Set up a curing room. Creating your own curing room is possible with the help of a brilliant product called a *coolbot.* The coolbot turns window air conditioners into cooling machines.

Mold and making salamis or other dry-cured products

You want a good white mold to develop on the outside of your dried salami. This mold is called Penicillium and is safe to consume. The mold can be applied with commercial mold cultures (applied to the outside of the sausage after fermentation). Or you can inoculate your drying room with a well-molded salami (like one you might buy at the grocery store). This technique works very well and it should only take about a week or so for you to see the beginnings of spores forming on the surface of your salami.

White mold is the protector of your salami, and it will help during the drying process and impart a small amount of flavor.

Bryan Butler, from Salt & Time, Texas, has this to say about mold: "You want white mold. If you see yellow or green molds, you can wipe them off with a 50/50 solution of vinegar and water. Red mold is what you don't want to see. If you see red mold, remove the cured meat in question and discard."

You also need these things:

- ✔ Lugs (bus tubs or large, deep plastic containers) for rubbing salt cure on meat or curing.

- ✔ Large resealable plastic storage bags.

- ✔ Twine or plastic zip ties for keeping the muscle in a uniform shape. Just cinch up on the ties as the muscle loses mass (and moisture) during the drying process.

- ✔ Casings (natural or artificial) for stuffing whole muscles or salami-esque mixtures.

- ✔ Ham bags for hanging meat (the effect is the same as wrapping in cheesecloth, only easier).

Ensuring safe curing practices

Practicing safe curing techniques is very important; many variables and environmental influences can negatively affect your meat during curing and make the final product unsafe to eat. Meat curing is a science, and to do it safely, you need to consider a variety of things.

Monitoring temperature

You must be able to monitor temperature precisely when curing meat. For a sodium nitrite cure to work, the nitrite must be able to react with bacteria in the meat. Yes, you read that correctly: Bacteria must be present in the meat — which is why temperature is so crucial to this curing method. For the nitrite to work with the bacteria, the meat needs to be at a temperature higher than 36 degrees Fahrenheit (meat freezes at 36 degrees and slows or halts the process) but lower than 50 degrees Fahrenheit (at which point the meat can spoil because the bacteria get out of control).

Checking pH levels and nitrite amounts

Safe curing practices also require that you monitor pH levels in the meat as well as the amount of nitrates you use. Salt mixed with nitrates is a powerful and capable preserving combination, like the dynamic duo of cured meats.

The USDA requires that you monitor pH (acidic/alkaline level) during the fermentation process if you are selling meat for consumption; you can accomplish this via pH strips or a pH meter. Most professional curers use a pH meter and test for the correct pH level (5.3 or below) once during the fermentation process. After the correct level of pH has been reached, the product is at a safe level of acidity and this prevents harmful bacteria or foodborne illness from developing and can then be dried. One really good resource with more detailed information about fermentation is a book called *The Art of Making Fermented Sausages,* by Stanley and Adam Marianski. It is a book anyone who is interested in meat curing should have in his or her information arsenal.

Avoiding botulism

Botulism is a food poisoning that grows in improperly sterilized canned foods and processed meats, and is not to be taken lightly. You can't smell it or taste it, making it a silent killer. Nitrates and nitrites are the most effective tool in combating botulism, as is using good-quality meat, clean ingredients, and the right amount of salt. For more information on botulism, see Chapter 15.

Time for the cure

Two types of cure are commonly used in meat curing: Cure #1 and Cure #2. You need to know how to use them properly because they are not the same. In fact, you use each cure for different types of products. Here are the differences between them:

- ✔ **Cure #1 (pink salt):** Cure #1 contains salt and sodium nitrite only. In America, the mixture is 1 ounce of sodium nitrite to 1 pound of salt. The nitrite keeps the meat safe for a short period of time, maintains a nice red color in the meat, and produces that well-loved "cured" flavor. This cure is typically used in products that are made and then cooked and eaten quickly, like bacon or fresh sausages. It is also always used on products that will be smoked at low temperatures.

- ✔ **Cure #2 (Prague Powder #2):** Cure #2 is a mixture of salt, sodium nitrate, and sodium nitrite. In America, the mixture is 1 ounce of sodium nitrite and 0.64 ounces of sodium nitrate to 1 pound of salt. This cure is commonly used on items that are dry cured over an extended period of time, like salami or cured meats. The sodium nitrite breaks down over time and becomes sodium nitrate (starting the curing process), which then becomes nitric oxide, an oxidizing agent that protects the meat from botulism, cured meats most formidable foe. *Note:* The USDA does not allow any nitrates to be used in bacon curing.

Making your own dry cure

You can mix up a large batch of a dry cure to keep on hand. The basic dry cure is simply salt, sugar, and cure. Here is a dry cure recipe that makes enough cure to cure 27.5 pounds (12.5 kilos) of meat at a ratio of 6.2 percent of the total weight. Therefore, to determine how much cure you need, use this formula: number of pounds × 0.062. For example, if you want to cure 16 pounds, you need 0.992 ounces of cure ($16 \times 0.062 = 0.992$).

BASIC CURE RECIPE, from Craig Deihl:

> 500 grams salt
>
> 250 grams brown sugar
>
> 25 grams Cure #1

Fats absorb only a certain amount of salt and then no more. Muscles, however, continue to absorb salt, and too much time in cure makes them salty. One way to test whether something is ready is to press the meat with your thumb. You want it to be firm, not soft or spongy. If a recipe calls for a rinse and soak step (like in the following bacon recipe), don't skip this step because it keeps the meat from being too salty.

Making your own wet cure brine

Here's a basic wet cure brine recipe. This recipe makes enough brine for any cut that can be submerged in one gallon of water.

> ½ cup salt
>
> ½ cup dark brown sugar
>
> 1 tablespoon crush red pepper
>
> 3 tablespoons pink salt
>
> 1 gallon water

Here's a quick country ham recipe you can easily make: Bone out the ham (size doesn't matter) into individual muscles, brine it for 3 days in the wet curing brine recipe, and then smoke it (see the following section for smoking instructions) or roast it.

About sodium nitrites

Sodium nitrites were isolated in the early 1900s (before this, salt peter was used). They are safe for consumption when used properly, yet some curing enthusiasts choose not to use sodium nitrites because they affect the flavor of the meat, making "everything taste the same." Nevertheless, nitrites do enhance the shelf life, flavor, texture, and most importantly, the appearance of the meat. The nice, rosy, red color you find in cured meats is a result of nitrites. Traditional, salt-only cures arguably produce more a pure, meaty flavor, but the resulting color — dark or brown, not rosy — can be unpalatable, even though the meat is perfectly safe to eat.

Smoke 'Em If You Got 'Em

Smoking preserves meat by slowing down the spoilage of fats and reducing bacteria growth, thus increasing the longevity of the meat and enhancing the flavor as well as the appearance of the meat. It has also long been found to be an antioxidant; the surfaces of smoked meat are a hard place for bacteria to take hold of. In addition to any variety of meat cuts within pork, chicken, rabbit, lamb, and beef, a wide variety of other foods are wonderful smoked: cheese, vegetables, salt, eggs, and nuts.

Smoking involves cooking meats at a low temperature (under 200 degrees Fahrenheit) over a long period of time with smoke from selected types of wood, imparting unmistakable flavor. It's a fairly easy process:

1. **Prepare the meat by curing, seasoning, or marinating.**

2. **Place the meat in the smoker for a period of time.**

   Smoking time depends on the size of the cut and the type of meat (brisket, chicken breast, and so on).

3. **Keep an eye on the meat in the smoker.**

   Take a look at the meat every once in a while. If it's a fatty cut, make sure that the grease isn't catching on fire.

4. **After smoking, remove and enjoy!**

The next sections give you the details and tips on what you need to smoke meat.

Gathering (or building) your equipment

To smoke meat (or anything else for that matter), you need only a smoker, wood, and some meat. You can smoke in a barbecue grill or purchase a smoker at any variety of home improvement stores. They come in a range of prices and with a range of options, but a basic unit — one that's the right size for your needs and gets it smoking — is sufficient. You can also build your own smoker.

Building a smoker at home is fairly easy and inexpensive. Virtually any kind of enclosure that is made from environmentally safe materials will work, as long as the smoke can make contact with the surface of the meat. If you're handy and want to give it a try, you can consult any of several resources that include easy DIY instructions on how to build a smoker at home.

Choosing your wood chips

The best kind of woods to use for smoking are hardwoods from fruit or nut trees. Fruit trees produce a light and fragrant smoke, whereas nut trees produce wood that is generally "smokier" or stronger in flavor. You can also try mixing woods to create a signature smoke: Play around and find your favorite combination.

Following are some common types of wood used for smoking:

- ✔ **Apple:** Sweet, mild wood with a fruity taste.
- ✔ **Alder:** A truly unique flavor that is perfect for smoking fish. Popular for smoking salmon.
- ✔ **Cherry:** Mild and fruity. A good choice for poultry, fish, and ham.
- ✔ **Hickory:** Arguably the most popular. Excellent for ribs and red meats. Can be very strong, so use sparingly.
- ✔ **Maple:** A light, sweet taste that's well suited for poultry and ham.
- ✔ **Mesquite:** A well-loved, strong, distinct flavor, but can become overpowering. Ideal for larger cuts of meat that require longer smoking times but not advised for poultry or fish because of the intense smoke.

    Consider using mesquite in combination with a milder and more fragrant type of chip to balance out the flavor.

- ✔ **Oak:** Ideal for larger cuts that require longer smoking times. Known for a strong but not overwhelming smoke and Texans choice for smoked brisket.
- ✔ **Pecan:** A mild, fruity flavor. Burns at a lower temperature than most other barbecue woods and is good for brisket, pork roast or chops, fish, and poultry.

You can purchase a wide variety of wood chips, chunks, and pellets online or from home improvement stores. Some smoking enthusiasts enjoy using wood from old wine or whiskey barrels. Perhaps this inspires you to try soaking some wood chips in red wine. Have fun and be creative!

Smoking tips

So how many chips should you use? Depends on how intense you want the flavor. Start with fewer smoking chips or chunks than more because you can always add chips to create more smoke, but you can't take chips away. Several softball-sized chunks in a smoker or grill (on top of the coals, of course) will get the job done. As you experiment, you'll find the perfect amount for your taste and preferences.

Soaking the wood in water before smoking is unnecessary. If you're using large chunks of wood, the water won't penetrate deep enough to make a difference, and if you're using wood chips, soaking them in water will not keep them from lighting up on top of a direct fire.

Making bacon

A great way to try your hand at smoking is to make bacon. You'll need one pork belly (refer to Chapter 9 for butchery instructions). Then follow these steps:

1. **Trim the edges of the belly to create a nice uniform shape and remove any glands.**

2. **Pound the belly with a flat mallet to even out the thickness of the belly.**

   This step help you have maintain consistency in your cure.

3. **Cure the pork belly for 7 days, using the Basic Cure recipe.**

   Refer to the earlier section "Dry cure recipes." You can also add some crushed red pepper, if you like.

4. **After the curing time, wash the belly and allow the outside to dry before smoking.**

   Depending on the size and environment, several hours in the refrigerator should do it.

From Scraps to Elegant Dining: Pâté, Terrines, and More

Dishes like country pâtés, rillettes, terrines, galantines, meatballs, and pork pies are elegant yet simple foods. Crafted for centuries in farmhouse kitchens, these recipes are evidence of hundreds of years of parsimonious preservation presented as elegant cuisine. A common joke is that a terrine is just a meatloaf with a French accent!

These dishes put to use every bit left over from the butchery process, transforming scraps into stupendous dishes. They are creative ways of using any number of ingredients or flavors to create layer upon layer of fat, cured meats, vegetables, forcemeats (a mixture of chopped or ground meat), spices, gelées, and more.

Making a meat paste: Pâtés

Although you're probably familiar with the one definition of pâté — a meat paste commonly made of poultry liver (foie gras, chicken liver) — the term more accurately means pie, or savory pie. In traditional French cooking, a pâté is made of raw pastry crust and raw forcemeat that is then cooked; essentially it's a terrine (see the next section) wrapped in crust. If you make a hearty pâté dough from lard, reserved and rendered from the butchery process, you can bring this dish full circle.

Pâtés can be quite festive because of the attention that goes into decorating the crust before it is baked. During baking, the dough rises and the forcemeat shrinks, and after baking and the pie has been allowed to cool, *aspic,* a gelatinous meat sauce, is poured into holes in the dough to fill the empty space.

Creating scrumptious layers: Terrines

A terrine, in culinary terms, is a meat, fish, vegetable mixture that's cooked, cooled in its container, and then served in slices. The container in question, or the terrine, is a deep oval, oblong, or rectangular baking dish (see Figure 17-1).

Often, when people think of forcemeats, they think of meatloaf, but terrines are so much more than meatloaf. They can be made of seafood, layers of sliced cured meats (like bacon), forcemeats, blanched or puréed vegetables, spices, nuts, and more. After they cool, you can even add a final layer of seasoned lard or gelée (a gelled suspension of anything edible).

The rich flavor and complexity in terrines is a result of the signature layering. The individually marinated or seasoned ingredients create depth during cooking, and the flavors ripen once again while the terrine "sets."

A good quality terrine is a worthy investment for every enthusiastic home cook's repertoire; terrines, whether for dinner or desserts, are a versatile piece of cooking equipment.

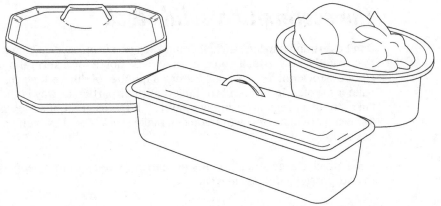

Figure 17-1:
A variety of terrines.

Illustration by Wiley, Composition Services Graphics

Upping the elegance factor: Galantines

Galantines are a dish of white meat or fish cooked, pressed, chilled, and served in *aspic,* a gelatinous sauce. These dishes are similar to terrines in that their preparation involves artfully layered meats, vegetables, nuts, or fish that are then pressed and covered in aspic. But instead of oven roasting them as you do terrines, the meat (or main elements) in galantines are poached in barely simmering stock, then cooled and ladled with aspic before being chilled.

Galantines can be as simple as a poached fish filet in a dish, seasoned and covered in aspic, or as extravagant as whole deboned birds (like chicken, duck, or turkey) or suckling pigs stuffed with forcemeats, poached, and then served cold. Galantines, though they are not commonplace on the dinner table, are quite spectacular, beautiful, and deliciously worth the extra effort.

Stocks and Sauces: It's All Gravy, Baby

While you're butchering, you'll end up with a healthy amount of bones. Don't throw them away; make stock instead! Stocks, soups, and sauces are a delicious way to use bones and trim while stretching your dollar. Stock is easy to make, and you can freeze it in ice cube trays and then store the frozen cubes in resealable plastic bags to use during cooking. A simple stock can become a beautiful soup, a rich sauce, or a gravy.

The secrets to a solid stock

Don't overthink making stock: It could keep you from ever making it at all, and that would be a waste. If you are going to be home for a several hours, put a pot of stock on. You can make stock a number of different ways, and most culinary professionals tend to think that their particular way is the right way. I'm more of a pragmatist: In my opinion if it tastes great, it's the right way. So following are just a few ways you can make stock; decide which you think is best.

You'll need a large stock pot, bones, carrots, celery, onion, peppercorns, thyme, parsley, and bay leaves.

You're probably thinking, "What? No measurements?" Nope, and here's why: Cooking is intuitive, and stock is one of those beautiful foods that shouldn't require a detailed recipe. Instead, use your instincts and common sense to build your own ratios of bone to herb to vegetable. For example, if you have 5 pounds of stock bones, simply add a few carrots, an onion, a few peppercorns, and so on. Really, it's easy!

✔ **Method 1: "The Slow Build":** Cover the bones with cold water (starting with cold water allows time for more flavor development), bring to a boil, and then reduce to a low simmer. Simmer the bones for several hours (3 or 4), adding more water as needed to keep the bones covered. After the bones have simmered for a few hours, add the vegetables, herbs, and peppercorns and simmer for 1 more hour. Season lightly with salt. Scrape any fat or scum that rises to the top of the stock pot every so often throughout the entire simmering process.

✔ **Method 2: "Roasty Bones":** Roast the bones on a sheet pan with a few sprigs of rosemary, shallots, and carrots until the bones are a nice, rich brown color. (I like to salt the vegetables and bones, too.) While the bones are roasting, cut an onion or two in half and sear one side of it (only one side) in a pan until it is blackened. Place the roasted bones, the roasted vegetables, the blackened onion, and the celery, peppercorns, thyme, parsley, and bay leaves in a stock pot and cover with water. (You can also add a couple tablespoons of tomato paste if you like.) Season lightly with salt and simmer for about 3 or 4 hours. Scrape any fat or scum that rises to the top of the stock pot every so often throughout the entire simmering process.

✔ **Method 3: "Just Put It in the Pot":** Cover the bones with cold water, along with the vegetables, herbs, and peppercorns. Bring everything to a boil and then reduce to a low simmer. Simmer for several hours (3 or 4). Season lightly with salt. Make sure to scrape any fat or scum that rises to the top of the stock pot every so often throughout the entire simmering process.

Whipping up a hearty sauce

Fear not! Even though many home cooks are intimidated by sauces, sauce making is actually very simple. Sauces can be made of pretty much any tasty liquid. They can be brothy, thick, olive-oil based, and so on. They can contain vegetables, herbs, vinegars, cream, yogurt, and more. And they can be served hot, at room temperature, warm, or cold. Really, the sky's the limit with sauces.

If the sheer magnitude of choices is what overwhelms you, don't think of everything a sauce can be. Instead, identify what your sauce is. For example, if you're making roasted pork chops for dinner and you want to make a sauce, what's available to you? If you have pan drippings left over from cooking the meat, or some leftover stock, start there. Use the pan drippings and stock as the base of your sauce. If you don't have stock, use water with the pan drippings.

Next, what flavors are complementary to what you are cooking? One flavor that goes well with pork chops, for example, is mustard; do you have any of that in the fridge? What about capers, lemon zest, marjoram, garlic, white wine, jalapeno, onion, fruit preserves . . . the list goes on. Choose a couple ingredients (don't overdo it) that work together and throw them in the pan. Then reduce the liquids (that is, let some of the liquid evaporate out) to intensify the flavor and let the flavors combine. Now, give your sauce a taste. Does it need anything? Salt? Pepper? Sugar? Chile flake? At this point, you can round out the flavors to create a balanced sauce.

Last, add your thickening agent. Here are a few easy choices:

- ✔ **Roux:** Mix together roughly equal amounts of butter (or other fat) and flour. Cook over low heat for several minutes. Then add this to your sauce mixture.

- ✔ **Beurre manié:** French for "handled butter," this thickening agent is just like roux except that you don't cook it. Simply mix equal amounts of flour and butter together (you can use your hands) and add it to the sauce.

- ✔ **Cornstarch:** Mix the cornstarch with water (or the broth from your sauce) and then slowly pour it into the sauce.

- ✔ **Butter:** Whisk butter into your sauce until you have reached the desired thickness.

Add the thickener to the sauce in the pan and let it thicken. When it's the consistency you like, pull it off the heat and pour it over your pork chop. Enjoy!

If your sauce *breaks* (separates), the balance of fat and water solids has broken, usually as a result of too much fat. To fix the problem, simply add a tablespoon or two of water and whisk the sauce back together.

Praise the Lard, Save the Fats

Fats and lard are so useful for so many purposes. They can be rendered for cooking oil and used in pie crusts, sausages, sauces, and even for soap making. A little chicken fat spread on toast with a fried egg and some greens is a fabulous way to start the day. Don't waste lards and fats; use them.

To render fat, follow these steps:

1. **Cut the fat into cubes and place in a large pot.**

   A slow cooker also works great.

2. **Put enough cold water in the pot to cover the bottom of the pot. Then place the pot on low heat, uncovered.**

3. **Cook until the liquid in the pot becomes clear, about 4 hours, stirring occasionally so that the fat doesn't stick to the sides of the pot.**

4. **Strain the fat and water through cheesecloth and discard any remaining pieces of fat.**

5. **Refrigerate the reserved liquid.**

   The fat and the water will separate, and the fat will rise to the top and become hard.

6. **Scoop the chilled fat resting on top of the water and — voila! — LARD!**

Part VI
The Part of Tens

The 5th Wave   By Rich Tennant

@RICHTENNANT

"I guess we should have expected this when he
named the chicken, 'Cordon Bleu.'"

In this part . . .

In this part, you have a chance to absorb condensed meat knowledge formed into a tasty, easy-to-digest format. Consider it pâté for your brain. The three chapters contained here have information that can help you avoid common (and sometimes dangerous) butchery mistakes, choose the best cuts for your grill, and use tricks the pros use to make great sausage.

Chapter 18

Top Ten Mistakes to Avoid When Butchering

In This Chapter

▶ Important rules

▶ Workspace advice

▶ Safety reminders

So much of butchery is about preparation. If you prepare your workspace, ingredients, and tools correctly, you'll be able to move smoothly and safely through the breakdown. But pare a hefty carcass with greasy fingers and a dulling knife, and the job becomes much harder and more dangerous. Throughout this book, I've included key prep and safety information. In this chapter, I tell how to avoid common pitfalls that lead to bad butchery, bad cuts, and bad backs.

Keeping a Messy Workspace

In butchery, as in surgery, the organization and cleanliness of your workspace has a direct effect on the outcome. Therefore, your tools must be clean and close by *before* you pull your meat out of the cooler. Do the following:

✔ Make sure you have an adequate cutting surface on your table. If you use a cut board, place a damp towel flat underneath it to keep the board from sliding.

✔ Have a receptacle for each of the following: waste, trim, and fat.

✔ Place your knives, saw, and sharpening steel on the table but out of the way of where you'll be working. If you'll be using a bone scraper and a hook, have those at the ready, too.

✔ Have a pile of towels at hand to clean greasy hands, wipe the meat free of bone debris, and keep your work area from becoming a disaster area.

Letting Your Meat Get Warm

During butchery, keep the temperature of your meat 41 degrees Fahrenheit or lower. Here's why:

- At 42 degrees, conditions become conducive to bacterial growth. Bacteria-ridden protein can make you sick, and it shortens the shelf-life of your finished product.

- Warmer internal temperatures compromise the quality of the meat's texture and flavor.

- Warm meat means flaccid muscles and slick fat, which both make cutting with any grace or precision a challenge.

Not Following the Separation of Time or Space Rule

One sanitation principle created by the United States Department of Agriculture (USDA) states that when butchering animals of a different species, you should do so either at different times (a separation of time) or in different areas (a separation of space). If you're butchering a chicken and then a piece of beef, for example, do the beef first (it has less microbial action). After you are done with the beef butchery, thoroughly clean your workspace and all of your tools before tackling the chicken. You may even choose to use a different cutting surface for different species.

If you're butchering different species at the same time (you're tackling one animal while your partner or spouse tackles another animal), you need two separate work surfaces, separate tools, separate everything.

Not Watching Your Posture

Many aging butchers grumble that butchery is a young person's game. Lifting heavy hindquarters, hauling boxes, and standing on your feet all day on an unforgiving surface can break you down. You can avoid feeling wrecked at the end of the day by following a few simple guidelines:

- **Cut on a surface that doesn't force you into an unnatural posture.** One table height tip is to measure from the floor to the crease of your wrist. If you are tall, many tables are too low to work at for an extended period of time. Also, make sure you have room to stand squarely in front of the butcher table.

✔ **Wear supportive shoes.** If you work on concrete, consider putting down a padded, non-slip floor mat.

✔ **Take the occasional break and stretch.** It helps to give your back a break, as well as clear your mind. Butchery requires concentrated work in the shoulders, as well as pressure on the knees and hips. To help these areas in particular, try these maneuvers:

- Stretch your arms overhead (put your knife down first!) and grab up at the air, alternating between hands.

- Bend over and touch your toes (or as close as you can get), alternating the weight of your body from one leg to the other.

- Grab your hands behind your back to release pressure in your chest.

- Take a short walk.

Improperly Storing Your Meat

Don't let the last step in the process destroy your beautiful meat. Storage is vital to flavor, shelf-life, and food safety. After you finish cutting, let the meat rest in refrigeration for an hour or so before packaging it. The meat will lose some blood, and you want to let that moisture seep out before you package the meat for storage.

Cryovac, or vacuum-sealed plastic bags, is the most reliable and efficient packaging. You can also use a sealable plastic storage bag or butcher paper. If you use a bag, make sure to remove all the extra air before you seal it. Here are some guidelines:

✔ Package your meat in the sizes and quantities you'll actually use. If you are a two-person household, put two steaks together. Using an extra bag is better than chipping apart frozen meat.

✔ Label everything with a date and a description. Even if you think you'll remember, you may not. A sharpie is an underrated butcher's tool.

✔ If you're packing anything with a sharp bone, like a lamb rack, buffer the edges of the bone with paper towel so they won't destroy the seal and let air in.

✔ Try not to store anything in deep freeze longer than six months. If the freezer you keep meat in gets opened frequently, use up the meat in less than six months.

Letting Your Knives Get Dull

A dull knife is an injury waiting to happen. When your blade isn't sharp, you are forced to use excessive pressure, which can lead to a slipup. Always have a sharpening steel at the ready and sharpen your knives frequently throughout the butchery process. Stopping to hone your blade should be a natural part of your rhythm.

Wasting Perfectly Useful Scrap

Almost every little morsel of an animal can be put to good use. Good butchers will tell you that their waste bin is barely touched. Glands and some connective tissue are about the only parts that won't serve you. Here a couple of ways to avoid waste:

✔ Know where you want to end up before you begin. Think about what you would like to prepare before you butcher because doing so may impact how you break down the animal. Remember, butchery is a means to an end.

✔ Don't let your creativity stop at the dinner table: Soap, dog treats, and candles are just a few things you can produce from trim and waste. Many reference guides to whole animal cookery are available that can help you get the greatest value and the greatest pleasure from your new skills.

Rushing through the Process

When you're butchering, you don't want to fight the clock. Accidents happen when you are tired and lose focus. So don't try to force too much into one day of cutting. In addition to the actual cutting (which admittedly will take you longer when you're just beginning), leave time for these activities:

✔ **Setting up and cleaning up:** These tasks can be time consuming, but they're crucial parts of butchery. Factor in an extra hour for your work time to accommodate them.

✔ **Taking care of your workspace while you butcher:** You need to take the occasional break to wipe down cutting surfaces, sharpen your knives, put meat into the cooler, and take the next section out of the cooler.

✔ **Taking breaks to rest and stretch:** Butchering is a physically tasking job, partly because you're often dealing with heavy, sometimes unwieldy pieces of meat, partly because you're standing in one position for long periods of time, and partly because you're making repetitive motions. Be sure to give yourself time to stretch before beginning and take breaks as you need throughout the process.

Being Careless or Distracted

The most important thing to remember as a butcher? Don't get hurt! Butchery is dangerous. Using a sharp, greasy knife can — and often does — cause an injury. Every butcher I know has had an injury. You can keep yourself safer by following a few guidelines.

✔ *Always* cut away from yourself.

✔ *Always* keep your knife sharp.

✔ *Never* keep your knives in your work area when you're not using them. Many butchers use a scabbard to hold their knives rather than set the knife on the cutting table. A scabbard helps you avoid stabbing yourself when you reach around to pick up a piece of meat.

✔ Use your hands when you can. Often a good tug with a boning hook or your hands can do the job just as well as a knife.

✔ Stay focused. If you are tired or distracted, an injury is often just around the corner. I have noticed that cuts usually happen at the very beginning or very end of a shift. In the morning, employees get harried by the daunting to-do list, and in the evening, they are tired and ready to quit. Cut when you can devote your full self to the job.

✔ Clean your knife handle occasionally. Meat grease means a slippery handle.

Being Fearful

Yes, butchery can be dangerous, but don't let the fear rule you. A skittish butcher is a lousy butcher. Always be respectful of the risks, but respond by being patient rather than petrified. When in doubt, move slowly. Regard the animal. Read the instructions from beginning to end. Try to visualize what you want to do before actually doing it. Don't hack. Don't rush. Know where your fingers are in relation to the blade at all times, and wear the right protective gear when necessary. Be one with the process. And remember: Every expert butcher was once a beginner.

Chapter 19

Top Ten Grilling Cuts

Cooking over an open flame is caveman-ancient. In fact, some folks speculate that the word *barbecue,* which means slowly cooking meat indirectly over an open flame, is most likely derived from the Caribbean Taino word for "sacred fire pit."

Americans, for their part, have taken to grilling with fetishistic fervor, tying manhood, summers, and patriotism all up in a "Master of the Grill" apron string. In this chapter I share the ten best cuts for grilling. Here you'll find a few tried-and-true grillable cuts, as well as some overlooked and undervalued bits that make great grilling meats. I also include some preparation suggestions. The choices and advice may surprise — and inspire — you!

Chicken — The Whole Thing, Every Last Part

For grilling an entire chicken, I suggest preparing the chicken carcass the brick style way:

1. **Take kitchen shears and cut down the backbone on both sides to remove it.**

2. **Pop out the sternum and press down on the ribs.**

3. **Marinate the chicken.**

    I usually use something as straightforward as olive oil, salt and pepper, and some fresh herbs.

4. Place the chicken skin-side down on a hot grill and put a foil-covered brick on top.

Be sure to tuck the wings into the body so that they don't get petrified.

Ribs, Any Kind

Look, any food this messy has got to be good. So what ribs are best? All of them! And you have so many to choose from. Some animals even have more than one set, with different characteristics.

Ribs tend to call for a sauce or marinade more than other cuts on this list because they have a lot of connective tissue that needs all the acidic help it can get to break down. Here are some pointers for beef, lamb, and pork ribs:

- **Beef short ribs:** If you are going for something as thick and gristly as a beef short rib, braise them before putting them on the grill.

- **Lamb ribs:** Little lamby ribs are a super cheap and often ignored cut that makes a great grilling meat. I like to slather these with a smoky Mexican hot sauce, like Cholula, and some olive oil before I introduce them to the coals.

- **Pork ribs:** Can't have a "good ribs for the grill" list without including pork ribs! Baby back ribs, spareribs, St. Louis style, or untrimmed — they are all redolent of porches and sunsets and add a fatty, meaty center to any outdoor meal. Pork ribs can be slathered in barbecue sauce and grilled, or parboiled with a little liquid smoke and then sauced and grilled. They can be smoked, roasted in foil on the grill — so many great ways to make ribs!

Hamburgers, That Glorious Staple

The Almighty Hamburger! It is not exactly a cut, but who can talk about great grilling meats without include ground beef? When you choose your hamburger, try to avoid pre-packaged, Styrofoam-tray mystery meat. Instead, get it freshly ground at the butcher. If the counter has ground beef that has the telltale pattern of a fresh grind, that's good enough.

If you consider yourself an afficianado, consider requesting a burger blend from your butcher, made to your taste. Here are some suggestions:

- Try a combo of aged short loin, brisket, and chuck.

- A 50-50 ratio of lamb and ground beef works wonderfully for flavor and fat.

- Avoid pork, unless you want a grilled meatball.

Burgers for the grill should be 6 ounces, about 1-inch thick, and seasoned with salt and pepper. For a rare burger, flip the burgers four times, leaving them on the grill about 30 seconds each time. If you like your meat medium rare, add a couple extra minutes on the grill. Add your favorite cheese on the final rest. And I am definitely biased toward a grilled bun.

Show Me Some Leg, Lamb

I love recommending lamb leg for the grill because it is only through a butcher's eye that you could see it in this way. Generally, leg of lamb is a roasting meat, a role for which it is sublimely suited. But if you prepare it properly, it makes a great grilling meat, too. Follow these steps:

1. **Seam out the leg into five individual muscles.**

2. **Salt and pepper each vigorously, and add a little olive oil and some rosemary and thyme.**

3. **Grill just at medium rare.**

   The lamb will continue to cook a bit as it cools. *Note:* More well done means more gaminess, so don't think you can cook the lamb flavor out of the lamb.

4. **Let the meat rest for a few minutes before slicing and serving.**

   You'll be thrilled with the results.

Flat Steaks and Their Three-Dimensional Flavor

Flat steaks are the three large-grained, flavor-intense, hard-working muscles also known as the flank, hanger, and skirt steak. Because these cuts aren't very large, they're usually prepared and grilled whole. Some suggestions for getting a succulent result:

- **Marinate, marinate, marinate!** These cuts need a good, long marinade soak to help break down the fibers; otherwise, they'll toughen up on you. Make sure your marinade has lots of acid (added as vinegar, citrus, and/or wine, for example).

- **Don't overcook.** Please, please don't overcook the grilling trinity. Medium rare or rare is the absolute way to go. If you like well-done beef, then find a different dancing partner.

Pork Chops — Brine and Shine

Chops from the pig's loin and rib are not the most flavorful on their own, but the grill smoke brings out a perfectly balanced taste. Although they can dry out, they are succulent if prepared correctly. Follow these suggestions:

- **When shopping:** Go for a 1¼-inch thick, bone-in chop. (If you want super thin chops, just be ready to cook them super fast and super hot.)

- **When preparing:** Brine them so that they soak up lots of moisture and flavor. You can also add a rub before they hit the grill, but make sure it doesn't have salt in it or you might be up in the middle of the night running for a glass of water.

- **When cooking:** Take the chops off the heat right before they're done and let them rest. (They'll finish cooking as they rest.) It's a crime to overcook chops. Worry more about livery-tasting meat than trichinosis.

Flat Iron, a Butchers' Discovery

The flat iron steak has been promoted heartily by the beef industry in recent years, because those folks started to think like butchers. The flat iron muscle is gorgeous — except for a thick seam of connective tissue running through it that will seize up your meat. Once you know how to remove the seam (head to Chapter 12 for instructions), you can enjoy the well-marbled, deeply flavored goodness of this cut.

Lamb Saratoga, a Treasure Seeker's Prize

If you can find the Saratoga, you'll love it. This cut is as delicate as a tenderloin and has the deep flavor of the shoulder. If you love lamb, this is an excellent cut and can bear up to a medium-rare grilling. So why's this cut so hard to find? Mainly because of the regionality and confusion in the butcher's lexicon. The Saratoga is a boneless portion cut out of the shoulder, but people will tell you the Saratoga resides in almost every primal of the lamb. When you shop, ask your butcher where her Saratoga comes from. If she says the shoulder, you've got a winner.

Strip Steak, America's Sweetheart

The strip steak is America's sweetheart steak, and it goes by many names: Kansas City steak, ambassador steak, and club steak to name just a few. Whatever it's called in your neck of the woods, the strip steak hits all the marks: marbling, flavor, and tenderness. It sits in a lightly exercised part of the short loin, so it doesn't get tough.

To prepare, let your steaks get to room temperature, salt and pepper them, and then slap them on a super hot grill. You can get fancier if you like, but you really don't need to. These steaks bring their own fanciness.

The Rib Eye — There, I've Said It

The bone-in rib eye is often called the cowboy steak, which is fitting. This decadently marbled cut can handle the licking flames of an open grill. It is macho and elegant, just like your iconic John Wayne-type cowboy. And a 2-inch-thick rib eye, seasoned with just salt and pepper, pretty much does everything you need meat to do. Pair it with a small salad and a glass of gutsy red wine, and you're set for a memorable meal. A rib eye is hard to get wrong. Get it right, and you're in paradise.

Chapter 20

Ten Sssshhhhausage-Making Secrets

In This Chapter

▶ Food safety advice

▶ Tricks to improve your technique

▶ Ways to get the best texture and flavor

Making sausage is actually pretty simple. But as with most straight-forward arts, being successful is all about the details. The secrets of sausage-stuffing come down to the little moves that you discover and perfect after repeating the steps time and again. Lucky for you, I've already arced the learning curve, so you get to start from the other side. As with all butchery projects, take the time to get a handle on what you need to do, set up your workstation, and assemble your ingredients before you begin. Once you're in the meaty, fatty trenches, having to run out for peppercorns or grab a tool from the other room is much more difficult.

Keep It Cool

During sausage-making, you absolutely *must* keep the meat and fat at a temperature of 35 degrees Fahrenheit or lower at all times. Keep a thermometer by your side and temp the meat often (I use a basic dial-top metal skewer thermometer). If you see that it's getting close to 35 degrees, pop it into the freezer for a few minutes. Partially frozen meat is perfectly fine for sausage-making.

The low temperature is imperative for two main reasons: First, ground meat, with its very high percentage of surface area, is just the right environment for bacteria to grow. Temperatures below 42 degrees Fahrenheit do not promote the growth of bacteria. (In addition to keeping the temp low, you also must be vigilant in keeping the area clean, the topic of the next section.) Second, cold meat and fat improve the texture and taste of your sausage. When fat gets warm, it smears, and your sausage goes from succulent to greasy.

Keep It Clean

Keeping your food preparation area clean is important in any cooking situation, but it's especially important when handling the two main ingredients in sausage: the meat and fat. As I mention in the preceding section, ground meat has a lot of surface area and is very susceptible to bacterial invasion. Keeping the meat and fat at or close to freezing temperature is one important way to inhibit bacteria's proliferation, but it isn't enough. You also need to disinfect your work surfaces, utensils, tools, equipment, and hands before you start, throughout the process, and after you are finished.

Use latex gloves. Make sure they're powder-free, though, so that you don't inadvertently add any unsavory seasoning to your sausage!

Keep Notes

Sausage comes down to the simple chemistry of taste and texture. And like any good lab experiment, you have to track your discoveries. If you think you'll remember something, odds are you won't. Write it down. Sausage-making is way too expensive and time-consuming to make the same mistake twice. Every true meathead I know has a binder of scrawled-upon recipes.

Grind It Right

Grinding — that is, running your meat and fat through a grinder — is the first step in mixing. Much like making a pie crust, the trick is to avoid over-mixing because you don't want your fat to smear. So get your meat to the right consistency and then let it be; you will have a chance to mix more after the grinding is done. I also mix in the spice before I grind, so that the grinding process does some of the blending for me and helps me avoid the urge to overmix.

The kind of grinder you use depends on how much sausage you're making and what equipment you have on hand. To make a small amount of sausage (less than 15 pounds), most home cooks go with a grinder attachment for their mixers. If that isn't an option, you can always go with a manual grinder. And if you're doing a very small amount, you can mince the meat and practice your knife skills (consider this a silver meat lining). Don't know much about grinders? Head to Chapter 14 for details and buying tips.

Get in the Mix

After you grind the meat and fat, you're ready to do some mixing. You can use a stand mixer or just put on some latex gloves and go to town. (I mix my sausage in a metal bowl.) Because I add my spices before the grind, this step is more about getting the texture and consistency right. If you add your spices after the grind, it's also about getting the spices mixed in evenly. Here are some tips and tricks:

- **Keep the meat cold throughout the mixing time.** If your meat is warming up (above 35 degrees) from being worked, put it in the freezer until it cools down again.

- **Add cold water (or even crushed ice) to your mixture.** Doing so improve the consistency of your sausage and helps the spices bloom and spread. Paprika, for example, can become gritty if you don't add some water to your mix. Water (or liquids) should make up about 10 percent of the total weight of your recipe.

- **Make sure that you don't mix to the point where the fat and meat are all mushed together.** In your sausage mix, you want *particle definition,* a fancy term that just means that the fat and meat are still separate entities. You *want* to see some white fat globules in the mixture.

Test the Texture and Taste

Sausage is about taste and texture, two things you can test pretty easily during the course of making your sausage.

- **The texture test:** After mixing the sausage, you're ready to do the texture test. Hold a small bit of the mixture and squeeze your hand into a fist, letting the meat ooze out through the hole between your thumb and forefinger (make the hole the approximate size of the stuffing horn you're using). Look for visual clues to how the meat will flow from the stuffer. It should be sticky and pasty (this is the "bind"), and the lean and fat should be distinct (you don't want a smear of white fat). The texture should be uniform and flow easily; it shouldn't break off or be too crumbly. Add small amounts of chilled water or crushed ice, if necessary.

  Next, check for viscosity by putting some paste between your thumb and other fingers and pulling them slowly apart. The meat should stretch a bit.

✔ **The taste test:** Fry up a bit of the mixture and evaluate it. Does it have enough salt? Does it taste the way you want? This is your chance to make adjustments to the mixture (and to make a note of any additions, of course). You can also wrap a little piece of sausage in plastic wrap and drop it into boiling water. This test tells you whether you have the right amount of fat for a cased sausage (the liquids will escape and you can see how they react to cooking).

Hone Your Stuffing Technique

If you're making link sausage, you're going to be working with a stuffer. You can add a stuffing attachment to your mixer, but if you're doing more than 10 pounds at a go, I suggest a vertical stuffer. They aren't too expensive, and they make the sausage-stuffing go so much more smoothly. Either way, make sure that your entire apparatus is sanitized and that you lubricate the horns well with water. (Your casings will also be wet because you have to soak them well before stuffing.)

One of the keys to successful stuffing is maintaining the right amount of tension. Once you find this tension "sweet spot," you can more easily fill the casing in a way that produces a nice, tight link. Although getting the right tension is more of a "feel" you develop over time, here are some tips that can help you achieve success more quickly:

✔ **Pack the sausage mixture tightly to reduce the amount of air in it.** If you use a vertical stuffer, form the sausage into balls the approximate size of the opening (pack them so that you don't have cracks or air pockets) and then *forcefully* slam dunk these balls into the hopper. Don't miss, or you'll have a heck of a mess! Then pack down the meat again by hand. Repeat until the hopper is full. Slamming isn't mandatory, but it helps to settle the meat into the bottom. You can place the balls one by one and get a similar affect. Whichever technique you use, your goal is to prevent too much air from building up in the stuffer.

✔ **Don't let the links drop as they comes out of the hopper.** As the sausage comes out, support the casing with your free hand and gently lay the stuffed casing down.

✔ **Eliminate air pockets by pricking the links.** Air pockets cause your sausage to explode when you cook it. Prick wherever you see that any air is trapped. Squeezing to increase the pressure helps the air escape.

Practice Linking Tricks

In Chapter 15, I provide details on how to link sausage. Here are a few tricks that can help you avoid some common challenges you encounter when you're first learning:

- Don't overstuff the casing. Doing so can cause your sausage to burst when you're making links.

- Don't spin links in the same direction if you are planning on hanging more than three or four sausages down to dry.

- Always start with a tight square knot or bubble knot.

Store the Sausage Properly

Making sausage is a bit of a project, if you get my drift, and it's even more so if you're stuffing sausage. For that reason, you may want to make a batch big enough to have some extra that you can hold on to for later or pass along to very grateful friends and neighbors. Here are some general sausage-storing guidelines; for specific USDA recommendations for storage times, refer to Chapter 15:

- Keep your sausage mixture refrigerated at all times after mixing.

- Wrap it tightly in a plastic container. You can also use large sealable plastic bags or vacuum seal it.

Use Quality Seasonings

If you're taking the time to butcher quality meat and make quality sausage, don't skimp on the ingredients that impart simple, pure flavor. Always use quality seasonings. One way to really ramp up a sausage flavor profile you like is to spend a little time finding the right dry seasonings. In addition, fresh paprika, fresh ground seeds, and peppercorns have very bright aromas. (The dried stuff you've been keeping around in the cupboard for a year probably isn't the best choice, though it will work in a pinch.)

Index

• M •

• Q •

• R •

Notes

Notes

EDUCATION, HISTORY & REFERENCE

978-0-7645-2498-1

978-0-470-46244-7

Also available:
- ✔Algebra For Dummies 978-0-7645-5325-7
- ✔Art History For Dummies 978-0-470-09910-0
- ✔Chemistry For Dummies 978-0-7645-5430-8
- ✔English Grammar For Dummies 978-0-470-54664-2
- ✔French All-in-One For Dummies 978-1-118-22815-9
- ✔Statistics For Dummies 978-0-7645-5423-0
- ✔World History For Dummies 978-0-470-44654-6

FOOD, HOME, & MUSIC

978-1-118-11554-1

978-1-118-28872-6

Also available:
- ✔30-Minute Meals For Dummies 978-0-7645-2589-6
- ✔Bartending For Dummies 978-0-470-63312-0
- ✔Brain Games For Dummies 978-0-470-37378-1
- ✔Cheese For Dummies 978-1-118-09939-1
- ✔Cooking Basics For Dummies 978-0-470-91388-8
- ✔Gluten-Free Cooking For Dummies 978-1-118-39644-5
- ✔Home Improvement All-in-One Desk Reference For Dummies 978-0-7645-5680-7
- ✔Home Winemaking For Dummies 978-0-470-67895-4
- ✔Ukulele For Dummies 978-0-470-97799-6

GARDENING

978-0-470-58161-2

978-0-470-57705-9

Also available:
- ✔Gardening Basics For Dummies 978-0-470-03749-2
- ✔Organic Gardening For Dummies 978-0-470-43067-5
- ✔Sustainable Landscaping For Dummies 978-0-470-41149-0
- ✔Vegetable Gardening For Dummies 978-0-470-49870-5

GREEN/SUSTAINABLE

978-0-470-59896-2

978-0-470-59678-4

Also available:
- Alternative Energy For Dummies 978-0-470-43062-0
- Energy Efficient Homes For Dummies 978-0-470-37602-7
- Global Warming For Dummies 978-0-470-84098-6
- Green Building & Remodelling For Dummies 978-0-470-17559-0
- Green Cleaning For Dummies 978-0-470-39106-8
- Green Your Home All-in-One For Dummies 978-0-470-59678-4
- Wind Power Your Home For Dummies 978-0-470-49637-4

HEALTH & SELF-HELP

978-0-471-77383-2

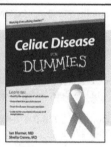

978-0-470-16036-7

Also available:
- Body Language For Dummies 978-0-470-51291-3
- Borderline Personality Disorder For Dummies 978-0-470-46653-7
- Breast Cancer For Dummies 978-0-7645-2482-0
- Cognitive Behavioural Therapy For Dummies 978-0-470-66541-1
- Emotional Intelligence For Dummies 978-0-470-15732-9
- Healthy Aging For Dummies 978-0-470-14975-1
- Neuro-linguistic Programming For Dummies 978-0-470-66543-5
- Understanding Autism For Dummies 978-0-7645-2547-6

HOBBIES & CRAFTS

978-0-470-28747-7

978-1-118-01695-4

Also available:
- Bridge For Dummies 978-1-118-20574-7
- Crochet Patterns For Dummies 97-0-470-04555-8
- Digital Photography For Dummies 978-1-118-09203-3
- Jewelry Making & Beading Designs For Dummies 978-0-470-29112-2
- Knitting Patterns For Dummies 978-0-470-04556-5
- Oil Painting For Dummies 978-0-470-18230-7
- Quilting For Dummies 978-0-7645-9799-2
- Sewing For Dummies 978-0-7645-6847-3
- Word Searches For Dummies 978-0-470-45366-7

HOME & BUSINESS COMPUTER BASICS

978-1-118-13461-0

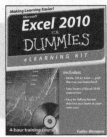

978-1-118-11079-9

Also available:
- Office 2010 All-in-One Desk Reference For Dummies 978-0-470-49748-7
- Pay Per Click Search Engine Marketing For Dummies 978-0-471-75494-7
- Search Engine Marketing For Dummies 978-0-471-97998-2
- Web Analytics For Dummies 978-0-470-09824-0
- Word 2010 For Dummies 978-0-470-48772-3

INTERNET & DIGITAL MEDIA

978-1-118-32800-2

978-1-118-38318-6

Also available:
- Blogging For Dummies 978-1-118-15194-5
- Digital Photography For Seniors For Dummies 978-0-470-44417-7
- Facebook For Dummies 978-1-118-09562-1
- LinkedIn For Dummies 978-0-470-94854-5
- Mom Blogging For Dummies 978-1-118-03843-7
- The Internet For Dummies 978-0-470-12174-0
- Twitter For Dummies 978-0-470-76879-2
- YouTube For Dummies 978-0-470-14925-6

MACINTOSH

978-0-470-87868-2

978-1118-49823-1

Also available:
- iMac For Dummies 978-0-470-20271-5
- iPod Touch For Dummies 978-1-118-12960-9
- iPod & iTunes For Dummies 978-1-118-50864-0
- MacBook For Dummies 978-1-11820920-2
- Macs For Seniors For Dummies 978-1-11819684-7
- Mac OS X Lion All-in-One For Dummies 978-1-118-02206-1

PETS

978-0-470-60029-0

978-0-7645-5267-0

Also available:
- Cats For Dummies 978-0-7645-5275-5
- Ferrets For Dummies 978-0-470-13943-1
- Horses For Dummies 978-0-7645-9797-8
- Kittens For Dummies 978-0-7645-4150-6
- Puppies For Dummies 978-1-118-11755-2

SPORTS & FITNESS

978-0-470-88279-5

978-1-118-01261-1

Also available:
- Exercise Balls For Dummies 978-0-7645-5623-4
- Coaching Volleyball For Dummies 978-0-470-46469-4
- Curling For Dummies 978-0-470-83828-0
- Fitness For Dummies 978-0-7645-7851-9
- Lacrosse For Dummies 978-0-470-73855-9
- Mixed Martial Arts For Dummies 978-0-470-39071-9
- Sports Psychology For Dummies 978-0-470-67659-2
- Ten Minute Tone-Ups For Dummies 978-0-7645-7207-4
- Wilderness Survival For Dummies 978-0-470-45306-3
- Wrestling For Dummies 978-1-118-11797-2
- Yoga with Weights For Dummies 978-0-471-74937-0